Radio Soundings

Zulu radio in South Africa is one of the most far-reaching and influential media in the region, currently attracting around 7.7 million listeners daily. While the public and political role of radio is well established, what is less understood is how radio has shaped culture by allowing listeners to negotiate modern identities and fast-changing lifestyles. Liz Gunner explores how understandings of the self, family, and social roles were shaped through this medium of voice and mediated sound. Radio was the unseen literature of the auditory, the drama of the airwaves, and thus became a conduit for many talents pushed aside by apartheid repression. Besides Winnie Mahlangu and K. E. Masinga, among others, the exiles Lewis Nkosi and Bloke Modisane made a network of identities and conversations that linked London and Africa with the heart of Harlem and the American South. As such, this dense network drew together the threads of activism and creativity from both black America and the African continent at a critical moment of late empire.

LIZ GUNNER is visiting research professor in the School of Languages, University of Johannesburg. She has taught in South Africa, Sierra Leone, and England. Her research primarily focuses on radio, popular culture, orality, and the intersection of performance and the political in Southern Africa. She has published extensively in journals such as *African Affairs*, *Research in African Literatures*, and *Journal of Southern African Studies*. Her most recent books include the co-edited *Radio in Africa: publics, cultures, communities* (with Dina Ligaga and Dumisani Moyo, 2011) and *Power, Marginality and African Oral Literature* (with Graham Furniss, 2008).

The International African Library

The International African Library is a major monograph series from the International African Institute. Theoretically informed ethnographies, and studies of social relations 'on the ground' which are sensitive to local cultural forms, have long been central to the Institute's publications programme. The IAL maintains this strength and extends it into new areas of contemporary concern, both practical and intellectual. It includes works focused on the linkages between local, national, and global levels of society; writings on political economy and power; studies at the interface of the socio-cultural and the environmental; analyses of the roles of religion, cosmology, and ritual in social organisation; and historical studies, especially those of a social, cultural, or interdisciplinary character.

For a list of titles published in the series, please see the end of the book.

Radio Soundings

South Africa and the Black Modern

Liz Gunner

University of Johannesburg

International African Institute, London
and
CAMBRIDGE
UNIVERSITY PRESS

CAMBRIDGE
UNIVERSITY PRESS

University Printing House, Cambridge CB2 8BS, United Kingdom

One Liberty Plaza, 20th Floor, New York, NY 10006, USA

477 Williamstown Road, Port Melbourne, VIC 3207, Australia

314–321, 3rd Floor, Plot 3, Splendor Forum, Jasola District Centre, New Delhi – 110025, India

79 Anson Road, #06–04/06, Singapore 079906

Cambridge University Press is part of the University of Cambridge.

It furthers the University's mission by disseminating knowledge in the pursuit of education, learning, and research at the highest international levels of excellence.

www.cambridge.org
Information on this title: www.cambridge.org/9781108470643
DOI: 10.1017/9781108556903

First published 2019

Printed and bound in Great Britain by Clays Ltd, Elcograf S.p.A

A catalogue record for this publication is available from the British Library.

ISBN 978-1-108-47064-3 Hardback

To Wiseman Masango, and to the memory of
'the King' K. E. Masinga

Be not afeard, the isle is full of noises
Sounds, and sweet airs, that give delight and hurt not.
Sometimes a thousand twangling instruments
Will hum about mine ears, and sometime voices,
That, if I then had waked after long sleep,
Will make me sleep again: and then, in dreaming,
The clouds methought would open and show riches
Ready to drop upon me, that, when I waked,
I cried to dream again. Caliban in Shakespeare's *The Tempest*, Act 3, Scene 2

Contents

Figures

Acknowledgements

There are a number of people who need a great deal of thanks in the making of this book. It began its life mainly as papers for seminars or workshops or conferences, but also from conversations, or even overheard remarks. The workshops or panels were often stimulating and nerve-racking; among these was the panel on radio at the African Studies Association of the UK conference in Leeds in September 2012, when Harri Englund, Jendele Hungbo, and I spoke on radio; another was the African Studies Association meeting in Philadelphia, 2013, and a panel on media and popular culture where I was interrogated, politely, by Hlonipha Mokoena and also met Peter Bloom.

I benefited hugely from contact with a huge range of people. And from contact with – through listening to or simply seeing – many different radios, usually quite small and often in unexpected places. One was the battery-operated radio used, among other things, for listening to serial dramas that sat in the bedroom of Mlomo Mhlongo, who ran a bar on the bend in the road in Qwayinduku, looking out towards the hills of the Ongoye Forest; another was the small transistor that would be hung up in the branches of the guava tree in the yard of the Dindi family, also in Qwayinduku, so we could sit and listen to it; another was the small, rather battered radio in the shop in the Shembe village of Ebuhleni on the edge of Inanda, Durban. There were others.

In terms of funding, I was fortunate to be funded by the National Research Foundation (NRF) for a project on 'Radio and the Making of Community in South Africa' while I was at the University of KwaZulu-Natal, Pietermaritzburg campus, and for a while thereafter when I was at Wiser, University of the Witwatersrand. This enabled me to employ Wiseman Masango, my valued research assistant, with whom I worked for a number of years. His extraordinarily detailed knowledge of Radio Zulu/Ukhozi FM, its announcers, and the serial radio dramas never ceased to amaze me. He is a kind of 'walking bible' on the topic. The funds from the NRF grant also enabled me to visit several archives, chief among these being the Sound Archives at the SABC in Durban, Old Fort Road, now K. E. Masinga Road, and also the SABC Sound Archives in Auckland Park. At the Durban SABC, Selby Goba, the radio tapes archivist, assisted me generously and, of course, very patiently. I was in awe of the resources that he guarded and ordered meticulously. He was indeed the keeper of treasures. After he retired I worked with Dennis Mseleku, who has also been unfailingly generous with his time, and with his expert knowledge of a range of subjects relating to the huge resource of radio dramas so carefully preserved by the Durban branch of the SABC. I cannot thank them both enough for their help.

Researchers can eat into the time of busy people, particularly broadcasters, so I was very fortunate to be able to interview such Durban SABC luminaries as Khaba Mkhize, Bheki Msane, Dumisani Nkosi, and Patrick Masonto Buthelezi. All these have now passed on and the world is a poorer and rather greyer place without them. Lawrentia Madlala and Winnie Mahlangu, two of the most famous voices in the long history of Radio Zulu and Ukhozi FM radio drama, also made time to talk to me, and I am very grateful to them. At the SABC in Auckland Park, I was able to interview Koos Hadebe, the famed football commentator and sometime radio drama actor. At the Johannesburg Sound Archives, where the main databases of all the variously situated regional archives are held, I was extremely fortunate to be helped by 'Uncle' Peter Raseroka on many occasions over a number of years. Ilse Assman, formerly at the SABC Sound Archives and then librarian at the SABC, was of immense help and a mine of information. Benny Jacobs, now Head of Sound Archives, was also extremely generous. My thanks to you all.

The grant from the NRF and research funds from the University of KwaZulu-Natal also meant that we could have a number of seminars and workshops on what seemed to be key areas that needed opening up in the study of radio beyond the ground-breaking work that had already been done by Keyan and Ruth Tomaselli, for instance. The focus on the cultures of radio, and on radio *in* culture, led to an exploration of popular culture, radio, and the media more broadly. A workshop in the early 2000s at the University of KwaZulu-Natal in Pietermaritzburg brought together Birgit Meyer, Joyce Nyairo, and Isabel Hofmeyr as well as colleagues and graduate students from the Maritzburg campus such as Khaya Gqibitole and Mandla Maphumulo. A later workshop, also on the Maritzburg campus, was conducted largely through the medium of isiZulu; on that occasion, among those present and contributing were Welcome Msomi, Eric Ngcobo, Lawrentia Madlala, and also Nakanjani Sibiya, a radio dramatist and novelist. Dina Ligaga from Wits spoke on her work on Kenyan radio dramas, then at an early stage. These events, and a later one held at Wiser, Wits in 2007, were hugely stimulating forums for avid, passionate, and often very knowledgeable discussions and debates about radio and its role and place in the country and the region at the time. My Wiser colleagues were always rigorous and supportive – a great combination. And sometimes they were inspiring.

I thank them for all that. I also wish to thank my colleagues on the Volkswagen-funded research project 'Passages of Culture: Mediations of Culture in African Society'. This ran from 2008 to 2013, was headed by Till Forster of the University of Basel, and brought together colleagues from Switzerland, Germany, Cameroon, Nigeria, and South Africa. It was a very productive interchange with rich research results and forged long-lasting links between those who were involved in it, and it made Africa seem a smaller, closer place, and Europe too.

I decided quite early on in this project that I wished to include in the book a section on the two exiled South Africans Bloke Modisane and Lewis Nkosi as they were intimately connected with radio, although, through no choice of their own, they were located outside South Africa and unable to return in the apartheid era. I made use of research funding from the University of KwaZulu-Natal to visit the Written Archives of the BBC, based in Caversham, Reading, and also to make use of the National Sound Archives, located in the British Library in London. In addition to visiting these two locations, I followed up on information about Lewis Nkosi held in the Transcription Centre records at the Harry Ransom Humanities Research Center at the University of Austin, Texas. Bernth Lindfors helped in alerting me to the importance of the Transcription Centre and I am most grateful to him. I have only begun exploring these rich records, which have so much more to yield. I also used the resources of the National English Literary Museum (NELM) at Rhodes University, Grahamnstown to research their valuable material on Bloke Modisane, and his correspondence with Langston Hughes.

Colleagues at Columbia University – in particular Marcia Wright and Gayatri Spivak – hosted a seminar on Bloke Modisane. Much of value came out of it, including comments by Mark Sanders that were of great help to me. A later workshop at NYU as part of a project on 'Commodities and Empire' led to my paper on K. E. Masinga that now forms the first chapter of this book. It was a most stimulating event. Some of the chapters have had earlier lives as articles, and now, in revised or mangled or cannibalised form, have a place in this book. Sections of the Introduction and Chapter 7 are drawn from Gunner (2000a), Chapter 5 appears as it is in Gunner, Ligaga, and Moyo (2011); Chapters 3, 4, and 6 each had other lives. The journals in which early versions of these chapters appeared are: *Current Writing: Text and Reception in Southern Africa* 15 2 (2003): 49–62; *Social Dynamics* 36 2 (2010): 256–271; *Social Identities* 11 2 (2005); *Social Dynamics* 41 1 (2015): 124–139.

All those I mention, and many more, including my most patient readers, Graham Furniss and Kai Easton, and wise friend Loren Kruger, have helped in this project. My sincere thanks to you all. My recent colleagues at the University of Johannesburg in the School of Languages have encouraged me cheerfully and in some cases with a certain ruthlessness to complete this book. I am grateful to them. My thanks, too, to Marne Pienaar, Head of LanCSAL, and to my colleague Thabisile Adams, who checked the isiZulu orthography and some of my translations when she had little time to do so. Dennis Mseleku, Head of the SABC Sound Archives in Durban, was patient and wise, and we are still friends in spite of my persistent claims on his time. Thank you, Dennis! My deep gratitude to Di Stuart for shaping and copyediting the manuscript and to Bianca Nienaber, my wise and skilled research assistant at the School of Languages, University of Johannesburg, for help in a number of crucial ways.

I have been very lucky with the publishers and skilled team of copyeditor, typesetter, and publicists at the IAL and CUP. I would particularly like to thank Stephanie Kitchen and Harri Englund from the IAL, the two anonymous readers, and also Judith Forshaw for being an amazing and very patient copyeditor. Veronica Klipp at Wits Press offered sound advice on many occasions and I am most grateful to her.

Finally, my thanks to Tom Gunner for many stays at his flat in Whitechapel when I spent time in the British Library.

Liz Gunner

Abbreviations and acronyms

ANC	African National Congress
ANCYL	African National Congress Youth League
AWC	African Writers Club
BBC	British Broadcasting Corporation
BBCWA	BBC Written Archives
BOSS	Bureau of State Security
BPCB	Bantu Programme Control Board
CALS	Centre for African Literary Studies
CBD	central business district
HAYCO	Hammarsdale Youth Congress
HRHRC	Harry Ransom Humanities Research Center
TC	Transcription Centre
IBA	Independent Broadcasting Authority
KADU	Kenyan African Democratic Union
KANU	Kenyan African National Union
MK	Umkhonto weSizwe
NAACP	National Association for the Advancement of Colored People
NAC	Native Affairs Commission
NAD	Native Affairs Department
NELM	National English Literary Museum
NET	National Educational Television (US)
NSA	National Sound Archives
PAC	Pan Africanist Congress
SABC	South African Broadcasting Corporation
SABC SA	South African Broadcasting Corporation Sound Archives
SADF	South African Defence Force
SAMRO	Southern African Music Rights Organisation
SWAPO	South West Africa People's Organisation
TRC	Truth and Reconciliation Commission
UDF	United Democratic Front
UNIA	United Negro Improvement Association
WDR	West German Radio

Introduction
Voice, race, radio

Sounds within: a visit, June 1999

As the radio dramas pressed on in the new dispensation within the South African Broadcasting Corporation (SABC), and in the country, Patrick Masonto Buthelezi, choirmaster and drama lover, younger brother of Alexius Buthelezi, continued in his post as head of drama at Ukhozi FM at the Durban SABC station, then still known as 100 Old Fort Road. This was the address to which listeners of the dramas (those produced from the Durban studios) were invited to send their comments, complaints, or requests at the end of every long serial drama. The visit in 1999 by Wiseman Masango and myself found Masonto Buthelezi seated in a large office with piles upon piles of documents lining the sides of the walls. They were, he informed us, radio plays sent in by eager writers who wished to find a space for themselves and their characters' voices in the prestigious drama slots of *imidlalo yomoya* (plays of the air). There was no evidence of any of the careful passing of memos between editors, including the vetting of scripts, stern requests for revisions, doubts, sometimes scorn, and so on, which, I would later discover, marked the era of BBC radio drama from the 1960s to the 1980s in which Bloke Modisane had a significant place. There was, however, an acute sense that Masonto Buthelezi knew what he was looking for in a good *umdlalo womoya*. And he also knew what he wanted.

Although there was no tangible evidence of listeners' views, there was a sense that, as he talked to us, they were present in his mind. The audiences of the dramas had, he said, an annoying habit of wanting a drama that they particularly liked or that seemed unresolved to continue to a second part. Buthelezi felt that the second parts of dramas requested in this way were rarely as successful as the original plays. Yet quite often he bowed to listeners' views and commissioned a second part. He was later to do just that with the riveting love story *Ebuhleni Nasebubini* (For Better and for Worse) by Dumisani Nkosi. So there was a sense, then, that even though a bureaucracy of drama production (so evident in the BBC files) was missing, in its place there was a constant knitting together of producer and receiver through the intimate passages of reception and reply that bound the community of listeners to those who brought the dramas into being, from Masonto Buthelezi himself, to the writers whose scripts were chosen, and to the producers, actors, and technical production team that together made the whole. Promising scripts that were considered not quite right were sent back by Masonto Buthelezi to their authors for further work, he told us.[1] New structures of feeling and the complex intimacies of writers, audience, and producers all combined to contribute to the mix of themes and treatments that were evident in the radio dramas of the closing years of the 1990s and into the early 2000s. Yet, the sense of continuity was also strong. Possibly what the dramas of the late 1990s did was provide a constant counterpoint to a number of possible ways of living. What was presented as established and known was often set against a serial drama crammed with turmoil and social confusion where there seemed to be no stability at all, only a menu of life choices that did not add up. Where was truth, the new dramas seemed to ask? Where was honesty, integrity, community? Where was pleasure? Where was the family? In some ways, as in the 27-episode serial drama by Mbuso Langa titled *Isiboshwa* (The Prisoner), the dilemmas of integrity versus wealth and corruption set out in quite an uncanny way what would be the national script a decade and a half later. Others captured a moment of life and then moved on.

Radio Soundings: South Africa and the Black Modern is in part about sound, radio sounds, and soundings. As Wiseman Masango and I sat in Masonto Buthelezi's office, with its piles of hopeful scripts, in a sense we were in a room full of sounds yet to be born, still to have a material presence within a serial radio drama. The scripts also contained a throng of voices still to be given material substance, a timbre, affect, a social life, and perhaps a death, through production within a play. In other words, we were in a room full of sounds. The somatic as well as the auditory nature of sound is a point well made by Charles Hirschkind in his profound study of ethics, the body, and social life linked to Islamic cassette sermons in Cairo (Hirschkind 2006: 8–9).

Voice

To link radio and sound, as this book does in its title, may seem banal and obvious. Yet there is a significant renewed critical interest in 'sound'. This has led to the theorising of political and cultural aspects of sound in relation to power and to global and colonial histories, interconnected with the local (Sterne 2003; 2011; 2012; Bloom 2014; Gautier 2014; see also Stoller 1984). Studies such as these put a different gloss on

[1] Interview with Patrick Masonto Buthelezi, 1999.

'the modernist story of the ear's decline' (Hirschkind 2006: 26–7, discussing Walter Benjamin's *The Storyteller* of 1969) and on 'a modern ocularcentric epistemology', which can be traced back to the late eighteenth-century thinking of the Enlightenment (ibid.: 13). All this makes the phenomenon of sound of key importance in any study of radio, a technology of the aural and the auditory rather than the visual and the ocular.

Because sound is unseen, its architectural and shaping qualities in the experiences of everyday life can be elusive. Sound can attune an individual to the dense lived texture of a modern city, such as downtown Cairo or the seething central business district (CBD) of Johannesburg. Repeated involvement as a listener in a particular kind of aural media, such as cassette sermons, so popular in many parts of the modern Islamic world, can hone particular affects and sensibilities. And with that 'tuning' can come an understanding of a moral code, a sense of where a person belongs or fits into a community. Sound can also have an intimate involvement with ethics, with the setting of bodily expressions and gestures that become part of a grammar of comportment linked to self-fashioning and identity (Hirschkind 2006: 8–9). Sound – and 'radio soundings', as the title suggests – is thus a concept to which the book attempts to give a particular meaning within the context of this study of radio, race, and identity in South Africa. The concept of voice is the particular frame within which I situate my radio analysis. The specific meanings of voice can shift: 'voice' can be set in a number of different registers, sometimes converging and sometimes remaining distinct, but central to this study are the ideas of giving voice and what it means to have a voice. These two questions are teased out by assessing the roles of particular broadcasters and writers – and, indeed, the radio dramas themselves – as mediating figures of voice. My text features a throng of characters, often multi-tasking, multivocal: unique programme hosts, DJs, drama writers, actors, and producers – they are all essential mediating figures. At once public celebrities and working artists of the radio voice, they have bridged the world of a dispersed listening community strung across the rural and urban spaces of the country, and the world enclosed in the dramas. Through an analysis of the radio drama genre within the broader theatre of Zulu-language radio, I ask what the mediatisation of the voice as sound meant in the context of the racially segregated South African state and its policies of black containment and subjugation. I then follow this question through into the post-1990 democratic era and further trace voice as part of the soundings of radio in isiZulu within a dramatically altered state and a differently convened public space.

As a setting for the above, I discuss, in brief, the wider political context of racial domination within which Zulu-language radio had its beginnings. I then complicate this picture by sketching the arena of black political thought and activism that existed contemporaneously within the segregationist and later 'apartheid' South African state of post-1948, when the Afrikaner nationalists first came to power. As I explore the idea of the black modern, part of my argument is that the long history of African-American entanglement with black South Africa, the mutual admiration, need, and succour of these two parts of the black world, became an essential component of black modern South African identity. In the main, through the radio soundings and the radio voices in this book, I attempt to pull together often submerged or fragmented stories and enhance understandings of hitherto perhaps half-perceived connections and narratives only half told.

In the book, I refer to orality as well as aurality. Both have a place. Recent studies have clarified the limitations of an approach to orality that sets it in the domain of the 'other' – in a 'natural' rather than a 'civilised' realm of order (see Sterne 2003; Bauman and Briggs 2003; Gautier 2014; Beecroft 2015). In the South African instance, orality – or orature, to adopt the term used by Ngugi wa Thiong'o – cannot be seen as 'premodern', a voice of alterity and the 'other'. Widely used to express identity, both private and public, and to mark kinship and respect, in many cases its forms were simply transferred to new uses and into new media beyond the unmediated voice. Orality and aurality each became part of the new voice in the public sphere and in private intimacies sometimes publicly voiced – finding a place not only in print culture in the columns of African-language newspapers such as the *Ilanga laseNatal* (The Natal Sun) but also in performance at political meetings of the African National Congress (ANC), where the alternative view of a demand for black political presence and power in governing South Africa made use of political forms such as praise poetry and invocations of powerful black rulers of the past. Orality, as it has played out in South Africa, thus assumes a powerful and ambivalent status. It can be seen at some points in history as a subject of the creation of alterity. But it can also be read as an invocation of an alternative political and social order. One of the arguments in this book is that orality through the technology of radio has fed into aurality and 'the mediatization of the voice as sound' (Fisher 2016: 2). The figure of the voice – to use Lauren Kunreuther's term – as it is played out in my analysis of a particular history of African-language radio takes many forms. Voice has shaped and been shaped by both urban and rural audiences. Voice also took on particular connotations: a range of affects largely linked – in the radio dramas that I follow as an emergent genre – to the family and also to conscience, ambition, greed, fear, and a range of social contexts. But within the frame of the family, the voice was 'the black voice'. This was a voice that carved itself out, often painfully and in the face of resistance, as a counter-voice. A phrase still frequently used in the Zulu-language print media is '*isizwe esimpisholo*' (the black nation) implying not a particular nation but a certain position, a certain consciousness. I argue that voice presents itself as a range of nuanced identities – both outside and within the dominant notion of the modern. Soundings thus points to a number of ways in which voice represents new subjectivities but also a new consciousness, and, in some cases, a defining of a new identity. Laura Kunreuther's study of radio voice and public intimacy in creating new modern democratic subjects in Nepal (2014) has echoes with the ways in which intimacy through radio dramas created new ways of negotiating a modern voice and self in the South African context. Daniel Fisher's studies of the mediatised voice in north Australian radio among scattered indigenous communities shows how voice can mark identity as a counter-presence to stereotypical settler concepts of indigenous Australians, and, in fact, give voice to transnational identities of blackness (2016: 9, 16). Here, too, there are links with this study: the radio voices I follow create connections and affects that resonate with a particular world of colour, and a modern identity with transnational lines of connection. In this study, we sometimes see the heightened politics around the right to have a voice, and around racialised voices. Voice gives presence to writer-actors as mediating figures, and, almost invariably, through the dramas the domestic sphere becomes the site of the listening environment; love, complications, and a host of individuals within or around the family take centre stage. To a large extent, the prism of the family provides the way into probing modernity, or possible modernities. We see how, by its imaginative remove, a dramatic genre enabled writers and listeners to claim a space for debate, as voices entered houses and rooms, or sometimes filled an open space and swung from a small tree, and people listened – and later discussed so that the capillaries of listening could spread through a community and a region. All this is part of radio voice.

Fisher has shown in his study how radio voice can be an index of power and a site of empowerment. Voice, Fisher argues, can bring about a 'profound suturing and repair of ruptured social relations' among modern Australian aboriginal communities (Fisher 2016: 20). In a far wider, global sense, it can simultaneously hold for its listeners 'a cosmopolitan identification of shared subjugation', so that they see themselves within the antipodean 'double consciousness' that was marked by W. E. B. Du Bois (1996 [1903]) and later by Paul Gilroy (1993) as key to black identity in North America and the Black Atlantic (Fisher 2016: 16). The function of radio voice to repair, reshape, and reinvigorate can thus operate in diffuse local networks, as in the north Australian case and more widely in indigenous radio in Australia. But, as Fisher points out, it also contains this wider consciousness of a shared subjugation reaching beyond the Australian continent. The present study tracks a different embodiment of radio voice, one located in the history of South African radio, race, and identity. Here, too, networks of identity and race and moments of shared history make it impossible to understand the specifics of South African radio without seeing its wider entanglement in the history of empire and of the African continent. The 'double consciousness' coined by Du Bois to capture the experience of the American 'Negro' had its own particular expression in the history of race in South Africa. This study maintains that radio – and, in particular, Zulu-language radio, on which I focus – played a key role in undoing, or at the very least deflecting, the heavy hand of race oppression by the Afrikaner nationalist government as it attempted to keep a great section of the South African population away from fulfilment, happiness, and achievement as modern subjects, and instead press them into a state of permanent servitude.

The political context

The wider political context within which African-language radio came into being has links with the entanglement of race with both modernity and identity. Radio also became a critical tool of modern technology for achieving the ambitions of those who were determined to govern the black population by domination rather than consent. 'Apartheid' was 'the electoral slogan which brought radical Afrikaner nationalism to power in South Africa in May 1948' (Dubow 2014: 1) and it became the context within which radio in African languages was forced to exist, and to find and maintain a voice, a sound. In many ways, the beginnings were not promising. But the power of the medium itself was not easy either to control or to contain.

One crucial date for Zulu-language radio is December 1941, when King Edward Masinga first broadcast the World War Two news bulletin in isiZulu from the Durban studios of the SABC, in downtown Aliwal Street. Possibly the first broadcasts by Masinga were not formally within the SABC system, but by September 1943 the news slot had a more permanent place (Mhlambi 2008; 2015). A second key date is June 1960, when the Afrikaner nationalist government, by then in power for 12 years, inaugurated with much fanfare the SABC station to be known as Radio Bantu. Within what wider context of black politics and black presence were Masinga's brief but quite revolutionary war broadcasts to be considered? Masinga, of course, would have been working from a carefully scripted wartime communiqué given to broadcasters by those monitoring war information. He could translate the script but change nothing. But he had his voice and he had 'radio sound', and with each of these he could confer on the Zulu language itself a new authority, even legitimacy. He did this with consummate flair. It was a kind of revolution (see Chapter 1).

Control of African movement and labour, masked by the benign term 'trusteeship', marked the early decades of white political rhetoric and policy after the formation of the Union in 1910. Laws governing black life proliferated. Most deadly was the sweeping Natives Land Act of 1913. It outlawed with the stroke of a pen various systems of share cropping that had meant black families could live on land now in the ownership of white farmers, and formerly, in some cases, their ancestral land. The upheavals it caused were dramatically and painfully documented by Sol Plaatje, journalist, linguist, politician, and first secretary general in 1912 of the South African Native National Congress, which later changed its name to the African National Congress (Plaatje 1916). Not only did the Land Act turn many thousands off the land, it also abolished the right of Africans to buy land outside the already restricted reserve areas. Blacks in white-ruled South Africa were indeed exiles at home.

In spite of the bureaucratic restrictions of passes and permits on free movement, blacks continued to leave the rural areas and seek some form of employment in towns and cities. Whole families moved, often living in whatever meagre shelter they could find (Muriel Horrell cited in Gerhart 1978: 28). In many cases, though, continuing the pattern forced on Africans by the hut and poll taxes, it was men who travelled long distances from rural areas to cities, especially Johannesburg, often recruited for work in the gold mines, where they lived in single-sex hostels and spent their working lives moving between '*ekhaya*' (home) and city. Many maintained the culture of building a homestead and a family in their rural area, despite their own long absences. Oswald Mtshali's dystopic poem 'Amagoduka' (Mtshali 1975: 128), for example, captures the harsh vibrancy of life between rural homes and the hostile environment of the mines.[2] Women, too, moved to the towns and cities, although in smaller numbers than men. Whereas in 1912 they made up only 19 per cent of all Africans in towns, by 1936 they comprised 36 per cent (Gerhart 1978: 27). There were many strands of black culture in this period. The new possibilities of modernity constantly absorbed into African languages, into poetry, song, and new vocabulary, meant that change shaped and reshaped the life of the everyday. And as advertisements in the black press show, there was a range of commodities and consumer goods that drew people into a changing world of taste, pleasure, and sensations. Alongside the harsh laws and often appalling living conditions there existed the capacity to make 'the new' and to create different modes of being. Networks abounded and cut across rural areas and towns: church, sport, political affiliations, horse racing, jazz, and football were all part of life. These social and organisational capillaries and arteries were activated and newly intensified through sound and voice on the airwaves.

[2] See Moodie and Ndatshe (1994) and Carton, Laband, and Sithole (2008) for important historical analyses of life in the mines. See Coplan (1995) on the trope of the migrant in the print literature and orature of the Southern African region.

The 1940s, the decade of Masinga's radio breakthrough, was momentous in many ways. It saw increased African urbanisation and more possibilities for African agency. Alongside drought and greater pressure on Africans living on white commercial farms, there were also more employment opportunities, particularly in the industrial Witwatersrand, and better-paid work outside mining employment. Africans migrated en masse to the cities, where they fought for their rights for subsistence, social life, and living space. All this put great pressure on the segregationist legislation that the South African state had painstakingly built up between World Wars One and Two (Shear 2012).

As World War Two began, the pressure to include South Africa's black population in the war effort against Germany through the use of the media must have been intense. The report of the Native Affairs Commission (NAC) of 1939–40, chaired by Deneys Reitz, covered the first 16 months of the war and was clear about the need for the communication of war news to 'the Natives':

> News: the official news bulletin issued daily in the Native languages by the NAD [Native Affairs Department] which is a summary of the press cables reaching South Africa freely from all parts of the world, has been accepted by the Natives as a reasonably accurate statement of events. (NAC 1941: 5)

The NAC report also covered points of tension with regard to the 'Native intelligentsia ... still thinking in terms of an assimilated citizenship with the Europeans', and mentioned in passing that Native trade unions were demanding recognition and stated firmly that they were 'a people just emerging from barbarism' (NAC 1941: 6–7).

It was in this atmosphere of national crisis and uncertainty, where the need to keep the population united behind the war effort coexisted with the belief in race separatism and the superiority of Western civilisation, that the first African attempt to breach the portals of the SABC was made (see Chapter 1).

In May 1942, with General Jan Smuts prime minister of a fragile pro-Allies government, Japan in the war against the Allies, support for Hitler from the Afrikaner organisation the Ossewabrandwag, and uncertainty over African loyalty, the hated pass laws, 'a fundamental instrument of white supremacist control', were relaxed in several major cities (Beinart 2005 [1994]: 130). The wider consequences of the entry of the Masinga voice into an African public space through radio could not yet be seen. His short war news bulletins read in isiZulu may at first have been followed by only a small number of listeners, although there was a huge African interest in the progress of the war. Black soldiers, although not allowed to carry arms – a point noted caustically in the black press – were seeing action in North Africa and in Europe. War news was covered by the black press, including in *Bantu World*, edited by Selope Thema, and by the Zulu-language Durban paper *Ilanga lase Natal*. The keen awareness of world currents of power and influence, ideals of justice and racial equality, which clashed with segregationist politics within South Africa's borders, grew sharply in the period 1939–45. Also present was the strong conviction among black leaders such as Dr Alfred Xuma and John Dube, editor of *Ilanga lase Natal*, of the importance of American Negro aspirations and the similarities of race oppression in America and South Africa. Both Xuma and Dube had been educated in America, Xuma in Europe as well (Gerhart 1978: 46).

Black political thought

There were also changes afoot in the wider world of black politics within South Africa. The newly invigorated ANC, led from 1940 by Alfred Xuma as its president general, was put on a sound financial footing with new organisational structures in place. Dynamic young men such as Anton Lembede, Jordan Ngubane, and A. P. Mda brought into the formerly staid 'trusteeship' political ideology of the ANC a new tranche of radical ideas. Lembede was elected first president of the ANC Youth League (ANCYL) in 1944 and was the 'principal architect of South Africa's first full-fledged ideology of African nationalism' (Gerhart 1978: 51).[3] His vision was pan-Africanist, and also, in part, that of a cultural nationalist. Although it would find echoes in the later writings of Kwame Nkrumah and Cheikh Anta Diop in the 1950s, it was ahead of its time for South Africa. Lembede also attempted to 'devise a formula for the mental emancipation of the oppressed' by emphasising race consciousness and self-reliance (ibid.: 65, 90). This was a search for the Africanist modern rather than moving on a path that linked progress with Western assimilation. The Lembedist 'rebel' mix of romantic nationalism and visionary pan-Africanism was generally not followed up by the ANCYL leaders after his early death in 1947. Yet, the infusion of energy and idealism in the ANC that came from the ANCYL in the mid-war years, combined with the increasingly repressive measures of the Afrikaner nationalist government under D. F. Malan, which came to power in 1948, led the ANC to move to extra-parliamentary tactics and mass mobilisation. In 1952, the ANC began a nationwide campaign of passive resistance which it called the Defiance Campaign. They began this two years after the Afrikaner nationalists put on the statute books the sweeping Suppression of Communism Act, the Group Areas Act, which zoned people according to race, and the Nazi-style Population Registration Act, which classified people by race. In the years leading up to 1960,

[3] For a fuller discussion on the Youth League, see Gerhart (1978: 44–83).

black resistance intensified and diversified, with Africanist radicals in the ANC breaking away in 1959 to form the Pan Africanist Congress (PAC), led by Robert Sobukwe. The PAC anti-pass campaign of 21 March 1960 led to the Sharpeville massacre and to the almost immediate banning by the Afrikaner nationalist government of both the ANC and the PAC. This was a legislative 'silencing' of the black political voice that would remain in place until 1990. It occurred less than three months before the birth of Radio Bantu in June 1960, when, ironically, a space was made by the same legislators for the mediatised 'Bantu voice'.

The black modern, race, and radio

In order to grasp the enormity of the failure to silence 'voice', and specifically radio voice, in the broader context of African life, we need to understand the importance of the long cultural and historical dynamic of African-American–African interconnection in South Africa. K. E. Masinga, Lewis Nkosi, and Bloke Modisane, who each feature in this study as specific radio actors in the narrative of black modernity (see Chapters 1, 3, and 4), were deeply involved in the art, politics, and educational strivings of African-Americans, and the impact on their own creative work and world vision is evident. This was not a one-sided engagement. Rather, it was a recognition of mutual need and respect. Masinga's 1950s translations for radio of nine Shakespeare plays earned him membership of the prestigious Mark Twain Society, a rare and valued honour repeated later in the decade when he was invited by the US state to lecture and tour in America. Modisane's long literary friendship with Langston Hughes saw the inclusion of his poetry in Hughes's 1960 anthology *An African Treasury*, and Lewis Nkosi, from his London radio base, drew on the acoustics, music, and rhetoric of the African south and of the black transatlantic. These three figures were playing their part, through radio, in a much larger conversation of blackness. For Nkosi and Modisane in particular, it was a triangular configuration, with London, the unpredictable heart of empire, as its centre point.

Generally, however, the important sources of the black modern are located in the long and enduring connection of African-American and African ties. These covered the fields of religious endeavour, expressive and popular culture, education, global black politics, and print. Robert Vinson argues convincingly that, as hopes of British intervention for racial equality and an end to segregation faded, '[m]any black South Africans glimpsed their modern futures not in the faces of Europeans but in the faces of American Negroes. The pathway to modernity was through African America, not a white Europe' (2012: 6).

Song and expressive culture more broadly were essential elements of the African-American presence in South Africa. The first performances of the Virginia Jubilee singers from the Hampton Institute in Virginia were in October 1890 in Cape Town. Their two visits and their tours throughout South Africa lasted for five years in total, had a huge effect on the morale of their African audiences, and made a lasting impact on black performance repertoires (Erlmann 1996; Vinson 2012). Their 'up from slavery' narrative of African-American success was performed through the combination of 'American Negro spirituals, dramatic comedic sketches and minstrelsy'. It held 'coded messages of exile and liberation, and made visible "a global colour line" more than a decade before the publication of W. E. B. Du Bois' *The Souls of Black Folks*', first published in 1903 (Vinson 2012: 13).

Also hugely important were the educational ties that saw key African leaders such as Reverend John Dube, and later Dr Alfred B. Xuma, investing in the educational resources built up by African-Americans at centres such as Booker T. Washington's Tuskegee Institute in Alabama. On his second trip to the United States, John Dube raised funds to begin his Durban-based Ohlange Institute, modelled on Washington's Tuskegee, lectured memorably at Tuskegee in 1897, and was ordained a minister of the Congregational Church. When Xuma returned to South Africa in 1927 after 14 years in North America and Scotland, his networks included key African-American missionaries such as Max Yergan, the actor, singer, and socialist Paul Robeson and his wife Eslanda, and prominent African-Americans such as Ralph Bunche and Ray Willein, a future executive director of the National Association for the Advancement of Colored People (NAACP), which was founded by W. E. B. Du Bois (Vinson 2012: 138). But these points of inspiration also extended to lecture tours by Xuma and his wife, Madie, throughout South Africa. Travelling extensively, their talks stressed the trials and lessons of American slavery, always emphasising that 'the Negroes themselves had taken an active part in making emancipation possible' (ibid.: 138).

The networks built up by the Xumas through lecture tours and through their performances in Johannesburg and Bloemfontein of a 1943 show put together by Madie Xuma and titled 'American Negro revue: the progress of a people' (Vinson 2012: 138) point again to the ways in which a particular vision of black modernity circulated as a counter-presence to a dominant discourse of white superiority. They also show a counter-presence in the face of a largely hostile or indifferent white press. There was, however, another singular factor in the formation of a counter-consciousness of race, progress, and black identity, namely the influence of Garveyism and its organisation, the UNIA (United Negro Improvement Association), in South Africa, particularly from 1920 to the late 1930s. Marcus Garvey's ideals of a united black world spread through local units of the UNIA set up in South African towns and cities as scattered as Cape Town, Kimberley, and Vrede, but also through his newspaper,

Negro World. Both became a means of constructing transnational racial identities, as well as formulating popular discontent. The paper and the ideas of the UNIA were circulated as early as 1920 by American Negro sailors as they passed through South African ports, particularly Cape Town. *Negro World* was sold in bookshops such as Jolly Jack Barnard's in downtown Commissioner Street, Johannesburg, which advertised regularly in Durban's *Ilanga lase Natal*. Until its demise in 1933, *Negro World* carried regular news of African politics, and it was certainly read by two of the key young radicals of the ANCYL, Anton Lembede and Jordan Ngubane. The Garveyist slogan of 'Africa for the Africans' was taken up and used by Lembede as first president of the ANCYL, in the early 1940s, to capture his 'Africans first' political position.

The ideas of a transnational blackness that circulated through the UNIA and *Negro World* cannot be seen as part of an elite black discourse but rather as something based on a determination to root out race oppression, which surfaced at all levels of society. The story of the success of Wellington Buthelezi, who masqueraded as an American Garveyist prophet figure in the Eastern Cape, inspiring the building of 'American schools' and bogus UNIA branches in the 1930s in rural areas such as Mount Fletcher, Tsolo, and Qumbu, shows that this popular consciousness of race oppression and longing for change also existed far from any urban centre (Vinson 2012).

Radio, the SABC, and the politics of culture

Radio has been a medium of huge importance in South African public life and in the production and negotiation of cultures in a complex, multilingual national space (Rosenthal 1974; Tomaselli, Tomaselli, and Muller 1989). Yet very little has been written on the place of African-language radio within this wider domain. One of the most far-reaching and important media in South Africa has been Zulu radio, which currently attracts an average daily audience of 7.7 million listeners.[4] Despite being such an influential medium in the country, there has been no book-length study of Zulu radio, known since 1994 as Ukhozi FM. *Radio Soundings: South Africa and the Black Modern* addresses the gap through its study of the station known first as Radio Zulu, which grew from Radio Bantu, set up in 1960 as the apartheid brainchild of the prime minister at the time, Hendrik Verwoerd. It explores questions about the role of language and forms of popular culture in the apartheid and post-apartheid eras and focuses on the genre of radio drama and its astonishing history over a period of more than 50 years. It asks

[4] See: http://www.ukhozifm.co.za/sabc/home/ukhozifm/aboutus (accessed 26 October 2018).

Figure 0.1 A family group relax after work with the radio in Vaalwater, Northern Province.
Source: Reinhardt Hartzenberg, AMO.

questions about the writers who created the dramas, and about those who produced and acted in them, and it explores the dramas themselves. What kinds of publics did the new genre create? What were the longer-term consequences in terms of the building of a bank of writers and listeners in an African language that, in the minds of the architects of apartheid, was to remain a rich site of the arcane and the exotic and nothing more. It was not to be part of the interactive 'civilised' modern world. How did the production of sound, of voices in intense dialogue grappling with issues of personal and social importance set in fragments of time, spliced with musical bridges, capture the passionate attention of a growing community of listeners over more than 50 years?

This book also follows the paths of two writers who were never within the SABC and its African wing, Radio Bantu. Instead, as exiles in a diasporic space beyond South Africa's borders, they chose to write for the BBC in London. Their dramas spoke, sometimes fiercely, to the idea of 'home', but they could also look beyond South Africa to engage with the black diasporas from Africa and the Middle Passage to the Americas and the Caribbean, and, more recently, to Europe. The new patterning of African and diasporic dialogue of which their work was a part can now be seen as a prefiguring of even wider global flows of connectivity.

If, at the end of the 1950s and into the 1960s, the exiled and diasporic South African writers Lewis Nkosi and William 'Bloke' Modisane found themselves part of a world in which the imagining of home fitted within a much wider redefining of place and community, so too were the writers of Zulu radio drama set in a broader network. Radio created the local and the parochial yet existed within a transnational arc, one in which programmes could be moved to other radio stations far beyond the borders of South Africa, where they reached diverse audiences and publics. Similarly, programmes that went onto the airwaves from within South Africa often came from other services such as the BBC or had their origins elsewhere in Europe. Thus, even the cordoned-off Radio Bantu was by default a segment of a wider media world. The world of the airwaves – and the internationally networked, interconnected organisation of the SABC – possessed an inherently unstable dynamism, a factor that made it as hard to control as it was seemingly malleable.

Control was the chimera that the apartheid media men and their overseers in government and the *Broederbond*, the powerful and secretive Afrikaner organisation that sought to promote its members at all levels of government and society, set out to gain. In some respects, such as the doctoring of news on the African-language sections of Radio Bantu, as on the English and Afrikaans stations, they were successful, although there were bold attempts to challenge misinformation. The Zulu Radio announcer Thokozani Nene always began his news readings by stating: 'The words I am about to speak are not my own.' For a time, he prefaced and ended his bulletins by declaiming the powerful and affective *izibongo* (praise names) of one or other of the Zulu kings, until he was instructed to stop, as this communicative act, which called up memories of an independent pre-conquest Zulu state, was considered subversive – which, indeed, was its intention (Scott 1999).

The need to use radio as a manipulative weapon in the battle to implement the untidy and somewhat ad hoc ideology of apartheid (Posel 1991) was often cloaked in the solemn and high-sounding language of aspirant nationhood. The political role of Radio Bantu within the SABC's programme policy was described by its director general, Douglas Fuchs, in 1969 as 'linked to national policy based on the recognition of the diversity of language groups' (Carver 1995: 81). Yet the multiple 'Bantu-language' channels were simply a part of the National Party's linguistic concoction of the different 'nations' or ethnicities within South Africa. Also crucial (see Chapter 5) was the party's deep fear of the modernising African voices within South Africa, and of the influence of the radio stations of the newly – or soon to be – independent countries to the north of the Limpopo (see Chapter 5).

It was in this treacherously unstable environment of the airwaves that the Afrikaner nationalist government, in 1960, with Hendrik Verwoerd as prime minister, began the experiment to control African-language audiences as part of its wider vision of racial containment and social engineering known as 'Separate Development'. The tensions and contradictions between an inward-looking policy driven increasingly by National Party ideologues intent on forcing radio to reflect its policies of Separate Development and the need to be part of a global network of broadcasting were never resolved by the SABC in the apartheid period. The contradictions were particularly sharp within Radio Bantu. The situation was inherently paradoxical: the effort to enforce ethnic purity and a static view of tradition was implemented through programmes set in the modern and modernising medium of radio, voiced through African languages that were, like all languages, not static structures but dynamic, shifting in response to 'social, historical and personal considerations' (Crystal 1987: 79).

The global nature of radio, which the SABC was often at pains to acknowledge, generally sat uneasily with the National Party's compartmentalising ethnic intentions for the African-language stations that constituted Radio Bantu. At particular moments, the transnational networks of the SABC could seem to chime with a particular Radio Bantu production – a case I explore in Chapter 2. Mostly, however, the African-language stations moved with their own programmes in a domain that was intended to be separate and othering. I argue that, ultimately, the grand vision for Radio Bantu – and for the SABC – failed. It was the very nature of the medium of radio that ensured this failure. Radio, in its production of new 'imagined communities', its far-reaching mediatisation of voice, and its range of aural communicative devices, enabled new audiences and publics to form. Previously unimaginable networks of communication came into being. Radio drama itself, which was to become an established genre of Zulu media culture, certainly opened up a space for the discourses of democracy and the making of the modern subject (Kunreuther 2014), even though it was – in part, at least – conceived as an instrument to be used, along with other Zulu-language programmes, to shape and control the mindset of the listeners. Designed to be merely a propaganda tool, Zulu radio became the sound of an aural African present rooted in diverse strands of popular culture and tradition and a new vernacular of the modern.

Black voices on air

Apart from a brief period in the mid-1920s when an African variety programme calling itself 'The Zulu Versatile Company' had a slot on Durban's radio station, the first programmes aimed at Zulu-speaking listeners came on the air in 1941 (Rosenthal 1974: 97). K. E. Masinga's riveting first Zulu-language war news bulletin was made on 23 December

1941. Only three minutes in length, it was nevertheless a clarion call of a future radio station that would develop against all odds as a counter-presence of having a voice and giving voice to shifting identities and subjectivities that could not be controlled or shaped by the stroke of a pen or by the sly words of an informer or controller. In his obituary broadcast (21 June 1990), his fellow announcer Hubert Sishi said that one of Masinga's favourite songs was by Frank Sinatra: 'I did it my way'. The track was played for '*UMangethe ogibel' amagagasi*' (Mangethe Who Rides the Waves), and the point well made was that Masinga, in an important way, had won his long battle for a station that spoke with its own voice.[5] Masinga was 86 and his long life had almost traversed an eventful century.

The dramas

The first play to go on the air was a musical script by the versatile and ambitious K. E. Masinga himself. Masinga – King Edward Masinga, as he became widely known – produced, in collaboration with Hugh Tracey, a radio version of a folk tale, *Chief above and Chief below*, with 'original songs by the author in traditional idiom' (Coplan 1985: 160; Tracey and Masinga 1945). Masinga was clearly a complex figure, seen by some as a sell-out and by others as a pioneer and a genius. He is possibly best read through the idiom of the folk tale (*inganekwane*) as a trickster character, a man of guile, wit, and tremendous energy and talent, moving between irreconcilable positions, and winning. The records show that contemporary radio drama (and musicals) in Zulu had a regular place on Zulu-language airtime from the early 1960s, and thus existed throughout the harsh apartheid years and into the new era, with its restructured broadcasting, and maintained its place in the new democratised structure.[6] *Chief above and Chief below*, the collaboration between Masinga and the musicologist Hugh Tracey, then director of the SABC Durban station, can be seen as heralding – as Birgit Meyer remarked of video films in Ghana – 'a newly evolving genre that has to be analysed as a distinct cultural form' (1999: 18). Radio drama – both in the earlier form of single plays or musicals and later as extended serials running over several weeks, and in some famous cases over many months – proved to be a genre capable of great flexibility and, through the buffer zone that the Zulu language itself provided, one that was shielded from the temptation of an easy mimicry or pastiche of English forms that writers in English such as H. I. E. Dhlomo had to contend with.

[5] SABC Sound Archives, K. E. Masinga obituary programme, 21 June 1990.
[6] SABC Sound Archives.

Due to the language employed, it was also a medium of expanding consciousness, allowing a wide engagement with forms of modernity and older forms within the cultural archive. The hybrid, visceral nature of Zulu greedily swallowed every shifting phenomenon and experience that the vibrant, harsh life of apartheid had to offer. Veit Erlmann's texts of some of Caluza's ragtime songs of the 1920s, such as the topical '*IXhegwana*' (Little Old Man), give a sense of this merging of musical form and also of the semantic scope, sharp social commentary, and easy use of neologisms in which urban Zulu was engaging during the 1920s in Durban (Coplan 1979: 139; Erlmann 1996: 77). The extent to which this linguistic richness was used and worked into the scripts of the dramas, and the extent to which the white controllers, loyal to the ideology of racial separatism and primitive purity, censored the use of 'the new', is hard to pin down. Certainly, as Khaya Gqibitole (2002) has shown with drama in isiXhosa, the heavy hand of censorship played its part, and both censorship and self-censorship were evident in the North Sotho or sePedi station (Lekgoathi 2009; 2011). Certain dramas, for instance, revelled in innovation while others from the same decade (the 1970s) appear, if read on one level, to ventriloquise a set position in their language and ideology (see Chapter 5). It is likely, however, that the scripts escaped some of the censorship that was undoubtedly imposed on news broadcasts and talks on current affairs, and even on popular music (Coplan 1979: 149). Moreover, as aesthetic forms and sound texts able to be read on a number of different levels, the full extent of these dramas' impact on both urban and rural listeners would have been difficult to gauge and to police. The scripts were thus voiced over the airwaves initially through the system of rediffusion at fixed centres, moving out to a growing listenership that included the migrant workers of the hostels, domestic workers, beer brewers, sex workers, part-time and full-time, and the growing number of white-collar workers, preachers, teachers, and nurses who constituted the small but growing urban elite. Also important in gauging radio drama's success is its links with topics of interest in the black press, and here love, courtship, and family life – those staple items of modernity – flourished as points of hot debate (see Thomas 2008: 31–57).

Establishing an audience, creating a community

Over the decades from 1960 to the 1990s, Radio Bantu, which was to become Radio Zulu and later Ukhozi FM, created a zone of public intimacy that must have been the envy of other radio stations. It built this up with listeners across urban and rural communities and in the creeping informal areas of settlement that were neither clearly one nor the other. It created new 'addressivities' and new speech genres. It was not tied to the local; rather, it may have begun a history of working in its own way within the linguistic and racial divides of the SABC, while also keeping alive long-standing transnational and diasporic connections. Its jazz and music programmes in particular were a means of keeping a finger on the pulse of local and global popular music and styles. It did this despite the heavy hand of the censor – which could scratch across the discs of Miriam Makeba and Hugh Masekela and ban a particular song because it referred to 'Tight Times' (Hamm 1995).

The early Zulu audience – which existed in Durban's compounds and factories where the first broadcasts were sent, by cable, and in communities such as the cluster of houses at Baumanville just behind the present SABC – was receptive, responsive, enthusiastic, and engaged. It was also demanding, and as it (initially) consisted for the most part of migrant workers and those who moved between the urban and rural areas, it constituted a community of listeners that was continually engaging with a number of different, often conflicting, kinds of cultural practices. Listeners of early radio were also often sophisticated practitioners of oral expressive forms – either through performance or through recordings – and were the audience for groups such as Solomon Linda's Evening Birds, Dem Darkies, and Mad Boys, whose music drew on a number of cultural idioms, both rural and urban, 'traditional' and 'modern' (Erlmann 1991: 68; La Hausse de Lalouvière 1996: 58).

Masinga and Tracey's first play, bringing into being a new aural space, can be seen as providing a point of entry into what Karin Barber has called the 'shifting, mobile, elusive space of the popular', which is itself continuous with both 'traditional' and 'modern' categories (1997: 7). Yet the peculiar social phenomenon of apartheid and the extravagant social engineering that it involved meant that radio drama in Zulu was forced to be a creature of multiple identities – sometimes gesturing to the popular and at other times relating to an emergent novelistic tradition closer to high culture than to popular culture. These two categories are themselves by no means mutually exclusive, and the boundaries between them are constantly on the move.

What was established early on in the radio drama was a kind of expectation that – like oral narrative – it would provide an arena where problems could be set out and some kind of journey towards resolution – or destruction – could take place. There was also the expectation of 'play' – that this was in some sense a free space outside the contingencies of daily life yet linked to them. What was being worked out were 'strategies of survival that made use of the best of both worlds, the urban and the rural' (Erlmann 1996: 168). This meant searching for new kinds of cultural expression that gave scope for wrestling with the present while still beckoning to the past in important ways. Thus, the genre of radio drama became a site where the complexities and ambivalences around 'tradition' and urban living could be debated and reviewed through the symbolic forms of the narratives of the dramas. There was no fixed 'tradition' any more than there was a fixed 'modern', nor was there a set vocabulary to define each state. Rather, a term such as *isintu* might be used to sum up an essence of tradition, and *isilungu* or *isimoden* for things outside some imagined site of tradition. Also – as in the lyrics of songs of the time by Caluza and others – the links between the rural and the urban space could be set up and the dramas could play their part in reorienting those who moved physically and psychically between the country and the city. Debates were made and sometimes decisions left open. It was a moving discussion with the dramas and their writer-actors as the key mediating voices. The role of these 'aural' forms for speakers of Zulu would have been similar to that of a magazine such as *Die Huisgenoot* in the earlier decade of the 1920s for its Afrikaans readers, speaking across the divide of town and country and 'recreating communities' for a mobile and disoriented people.[7]

From available records, the 1960s seem to have been the decade when serial radio dramas began to come into their own. Possibly this was because the creation of Radio Bantu in 1960, with its constituent language sections, meant that more time and money could be devoted to radio drama productions. It was also the beginning of the introduction of FM and the availability of cheap transistor radios, which were within the budgets of many poor households, both urban and rural. FM was a global happening as the technology around radio improved, but in South Africa its introduction was an ideological move. However, the aim of blocking out the reception of and access to certain stations was not always easy to implement (see Chapter 1). Other reasons for the strength of serial radio dramas through the 1960s, 1970s, and 1980s are likely to have included the concentration of skills and talent that this new aural genre was able to tap into. This may have been because other outlets were limited for black artists wishing to work in print media, in theatre drama, or even in popular music. David Coplan (1979) and Loren Kruger (1999) have both written of the harshness of the 1960s for African performers – musicians, singers, actors – with apartheid laws pertaining to mixed-race audiences, curfews for Africans, and so on making it hard for them to perform in public (Coplan 1979: 152). In addition, the 1960s was the decade of exile for many performers; most of the actors and musicians from the opera *King Kong* never returned. Peggy Phango, for instance, who played the lead in the 1961 London production, died in London in 1998 having made another career for herself in cabaret, and later on stage and in television in Britain (Knipe 1998). Miriam Makeba,

[7] See Beinart (2005 [1994]: 115).

Hugh Masekela, and Letta Mbulu were among the many who went into long exile.[8] Others, such as Can Themba and Nat Nakasa of the *Drum* school of writers, died in exile, Themba in Swaziland and Nakasa in New York.[9] They did, however, mark the black modern through their essays and short stories (see Patel 1985; Stein and Jacobson 1986). Here, too, mediatised voices and the black modern intertwine: Masinga's Shakespeare translations were broadcast in the 1950s on the English-language service of the SABC, and probably also on rediffusion cable radio. According to Masinga, the first, *The Merchant of Venice*, was very popular in towns and cities, and *Romeo and Juliet* 'took Johannesburg by storm'.[10] Shakespeare on the airwaves in isiZulu in the 1950s must have accounted for the famous taste of Sophiatown gangsters not only for American style dress, but for hearing Shakespeare recited on street corners (see Chapter 3), although this was in English.

In this harsh climate, Zulu radio drama was able to survive, clearing a space to voice, sometimes in disguised forms, the anxieties of people's lives, without ever touching overtly on the severe politics of the time. The 'journeying' aspect of the narrative structure was emphasised in the serial dramas that allowed the audience time for the problems and conflicts that were worked out in the episodes to run alongside the dramas of their own lives. The serials provided a combination of intimate counterpoint and meta-commentary, although, of course, there could be no direct allusion to apartheid.

The first serial drama listed in the Johannesburg SABC Sound Archives holdings is called *uDeliwe*; it was written by Mandla Sibiya and ran for 25 episodes in March 1964. It must have been enormously popular as it was followed in May by a sequel, *Khumbula uDeliwe* (Remember Deliwe), which ran for 59 episodes. In some cases, novels were serialised, and in December of the same year C. L. S. Nyembezi's *Inkinsela YaseMgungundlovu* (The Flashy Man of Pietermaritzburg), which had appeared in print just three years earlier, was serialised over 37 episodes. Clearly, by 1964, not only had a 'knowledgeable community' for radio drama been firmly established, there was also a group of writers, actors, and producers deeply involved in the production and propagation of the genre. Thus, radio drama – together with forms of popular music – became a way of maintaining a fragile bridge between classes and groups in the fragmented community of Zulu-language speakers. In this instance, the producers were the writers and

[8] See Makeba and Hall (1988).

[9] For an addition to the extensive writing on the *Drum* period, see Sylvester Stein (1999). For an angry comment on the book, see Harry Mchunu, 'Drink "not cause of Nakasa's end"', *Independent on Saturday*, 23 October 1999. See also Brown (2014).

[10] SABC Sound Archives, Masinga interview, uncatalogued.

intellectuals of the small black middle class, while the community of *listeners* constituted a much broader cross-section of urban and rural Zulu society. The drama thus provided a common space of enjoyment and for interrogation of forms of contemporary culture and politics, of the modernising process, and of the claims of 'tradition'.

Among the producers of these radio drama serials names such as Hubert Sishi, Joshua Mlaba, Alexius Buthelezi, and Bhekinkosi Ntuli (the eminent writer D. B. Z. Ntuli) appear time and again. Sometimes they feature as both author and producer, as in the case of Hubert Sishi, who joined the SABC in 1945, close on the heels of K. E. Masinga. *Isikhumba Sebhubesi* (The Lion's Skin) which he both wrote and produced ran for 80 episodes from February 1968, with numerous repeats in subsequent years. Winnie Mahlangu, who played in the musical *uNokhwezi*, went on to translate from the isiXhosa radio drama *Buzani kuBawo* the isiZulu version, *Buzani uBaba* (You'd Better Ask Father).[11] Here, patriarchy is scorchingly seen as causing the demise of a son's happiness due to a rigid and unchanging father who demands a traditional marriage and not one for love. Later, by the early 1980s, she had her own programme, *Akesishiyelane* (Let's Share Things Together). Particular actresses and actors began to appear frequently; one of them was the actress Lindiwe Ntuli, who was in the marathon *Isikhumba Sebhubesi*. The names of people later widely known in South African show business, such as Welcome Msomi (adapter and producer of *Mabatha* – the Zulu musical adaptation of *Macbeth*) appear in the cast lists; Msomi was also an author. Again, what is noticeable is the doubling up of roles. In March 1969, in one of the still short 30-minute radio dramas, Joshua Mlaba produced *Ngiyazi Mina!* (I Do Know!) and Welcome Msomi is listed as both writer and an actor. In May of the same year Mlaba produced a serial drama of 32 episodes, *uMaqinase* (Miss Headstrong), featuring Welcome Msomi and Lindiwe Ntuli among the cast. The space for performance, authorship, and production that radio drama in both its serial and single unit form offered allowed individuals to develop in some cases multiple skills that were fed back into the ongoing production of radio drama. It also meant that individuals such as Welcome Msomi were able to move out into the arena of stage performance and production. The 'strategies of survival' that Erlmann speaks of as having marked the making of culture by those who travelled between urban and rural centres and diverse cultures must also have been used in a different way by the producers of radio serials. In a time of cultural marginalisation and political oppression, those working with the spoken word and the voice in this aural form often became multi-skilled, and, through a kind of

[11] Interview with Winnie Mahlangu, 2006.

marginal, liminal existence, they were able to continue in an important area of cultural production and cultural brokering.

Hiding in the thicket of language

How did the radio dramas protect themselves from being overrun by the thought police of the African-language media in the years before 1990? Were its vitality and popularity linked to the restrictive years of apartheid? Had the genre, its makers, and its public worked out a means of – almost – bypassing the forced silence of those years? Possibly the thick medium of the language itself provided such a means of survival. There was a sense in which the ideology of apartheid did not penetrate the way in which the language operated as a carrier of multiple signs and discourses and as a medium for the transformation of consciousness. As Stuart Hall has put it in discussing language and ideology:

> Language in its widest sense is the vehicle of practical reasoning, calculation and consciousness, because of the ways in which certain meanings and references have been historically secured. But its cogency depends on the 'logics' which connect one proposition to another in a chain of connected meanings; where the social connotations and historical meanings are condensed and reverberate off one another ... these chains are never permanently secured, either in their internal systems or in terms of the social classes and groups to which they belong. (Hall quoted in Kuan-Hsing and Morley 1996: 40)

The 'multi-accentual' nature of language, as Voloshinov (quoted in Kuan-Hsing and Morley 1996: 23) has put it, meant that multiple fields of meaning could be played out in a medium such as radio – and particularly in a genre such as radio drama, which could work with the real and the fantastic and exploit multiple levels of signification. Voice also held secrets of intonation, timbre, and affect that were not easily read by some. Radio drama, the *imidlao yomoya*, like popular culture itself, also became the space in which transformations could be worked out (Hall quoted in Samuel 1981: 232). The multiplicity of meanings in the aural text could also allow a piece to create its own world, at an imaginative remove from experienced reality, and still resonate with the historical moment (see Chapter 6).

If, as Barber suggests (1997: 6), forms of popular culture are often about suffering and aspirations for a better life rather than fantasy, much of Zulu radio drama can be seen as falling within such a range of subjects. Like other genres such as *izibongo*, radio drama can be seen as part of a long cultural tradition, produced 'in order to be' and 'in the expectation of being interpreted'. It is also a form that displays all the eager energy of the hybrid and the creolising. A recognised genre with its own knowledgeable and active audience, it affects the 'durable inscription of a sign' (ibid.: 8).

Partly through the ongoing interaction between audience and scriptwriters, Zulu radio drama has become an important space where issues can be worked out; these include tensions in the patriarchal home and the ongoing friction between older and younger generations, a topic with a long history in Zulu culture.[12] The dramas, moving steadily often over a six-week period, and sometimes far longer, have become a way of establishing an ongoing conversation on such issues. The censorship of overtly political themes did not mean that what was being produced did not engage with the harsh consequences of apartheid, or with the desire of listeners to find a medium that would express both their suffering and their search for a better life.

I first listened to a radio drama in 1976 when I was living in Mthunzini in northern Natal. I remember a character called Mandla (Strength) who was young, loved, and good, and who was pursued and persecuted by an evil older man aided and abetted by an unpleasant older woman. I recall that Mandla triumphed over his persecutors in the end. In some ways, then, the pattern I remember from that time – of a struggle between people or groups of people who are very real but can also be read as metaphors for certain opposing conditions or forces – has not changed significantly. However, I would argue that within the wide oppositional narrative frame there is a great deal of room for manoeuvre as the characters move through their journey from one state to another, and that an enormous number of themes and conditions of life can be examined and a great many social nuances inserted. It may be that in the 1970s the audience listened to the dramas as well as being involved with other genres, which provided individuals with different ways of patterning their own experience and forming and reforming their own multi-layered sense of identity. Thus, the community of knowledgeable listeners that formed the audience for radio drama was often also the audience for the plays of Gibson Kente that toured South Africa's townships in the late 1960s and 1970s. Kente's plays, his three successful musicals – *Sikhalo*, *Lifa*, and *Zwi* – and the later, more political *How Long?* and *Too Late* all used a set of conventions that, to some extent, was quite similar to radio drama.[13] In Kente's plays, with their popular aesthetics and deep-seated and deeply felt moralistic content, a strong principle of oppositionality propelled the action, as Bheki Peterson has pointed out. Peterson mentions the plays' evocative melodramatic spectacle that is profoundly informed by township culture, its slang, fashions, topical issues, music, and dance styles. The egalitarian impulse of Kente's form is matched by his predisposition to focus on the day-to-day struggles of his

[12] See Ben Carton (1999; 1998).

[13] See the rich and detailed account of Kente's career from Robert Kavanagh (1986: 113–44) and the script of *Too Late* in Kavanagh (1981).

characters, who are drawn from ordinary people, depicting and celebrating the lives they have wrought from the hostile and degenerate urban landscapes. Politics is always present in Kente's work, but it is nothing more than a surface reality, an aberration that masks the fundamental moral conflicts that challenge individuals and society (Peterson 1994: 47–50).

Besides the popular theatre of Kente, which demanded its own set of responses and aesthetic fine-tuning from its largely urban audiences, another shared genre, particularly for those who moved between the urban workplace and a rural base, as many men and some women did, would have been *izibongo* (praise poetry). In its popular form, this genre focuses on compact, often highly metaphorical statements of personal identity and includes both social comment in the words of the individual *izibongo* and social memory through the dances that accompany the performance (Gunner and Gwala 1991: 36, 158–223). The families I met in Mthunzini in the early 1990s who listened to radio dramas were also deeply involved in *izibongo*, *amahubo*, and other genres of song that people use to mark significant events in their lives, particularly in rural or semi-rural areas.

What I am suggesting here is that we need to remind ourselves of the range of genres – both popular and traditional – that a single individual would have been actively involved with as a listener/receiver and possibly as a performer during a particular period. If one wants to take a long view of a genre such as radio drama, it needs to be seen as part of a constellation of cultural 'engagements' and involvement. It becomes part of the active cultural knowledge of an individual, knowledge that is both changing and constant. And the constellation of engagements within which radio drama sits will vary according to the class of the individual and their local and social identity – and the era.

But why should radio drama have kept and even increased its audiences over the 50-plus years of its existence? Reasons may include the long-standing stability of the interaction between producer and listener, and the audience's active engagement through contact with producers and scriptwriters. Thus, at the end of every drama, listeners are invited to write or phone with their comments. And, according to producer Bheki Msane and scriptwriter Dumisani Nkosi, they do so in huge numbers.[14] The dramas, then, are important in the listeners' daily lives and there is an intimate dialogue between producers and audience. This intimacy not only marks the dramas and their audiences; it is more general. Keyan and Ruth Tomaselli remark on the way in which letters form a close link between listeners and programme announcers/producers (Tomaselli, Tomaselli, and Muller 1989: 96). They refer to Les Switzer's study of media production and reception in the Ciskei (1985), which shows how important the

[14] Interview with Bheki Msane and Dumisani Nkosi, 1998.

isiXhosa programmes programmes were for both rural dwellers in the Ciskei and legislators, and they provide a quite astonishing statistic:

> It is worth noting that in 1981 the Bantu Language Services received 5,530,000 (five and a half million) letters from their listeners. (Tomaselli, Tomaselli, and Muller 1989: 96)

The Switzer evidence gives the impression that the African-language services were particularly strong in rural areas. Yet the sense gained from interviews with listeners to Zulu radio dramas conducted by Wiseman Masango in and near Pietermaritzburg, and from Bheki Msane's and Dumisani Nkosi's remarks to me at the Durban SABC (both in the late 1990s), is that urban dwellers also have a long history of listenership.

The chapters that follow engage both with the makers and making of radio drama and with the dramas themselves. They also focus on the two writers, Bloke Modisane and Lewis Nkosi, who used radio and radio drama in exile, working not with the SABC but with the BBC or with an 'independent' programme maker, the Transcription Centre. In essence, then, my text focuses on the fine grain of cultural production through 'soundscapes', and in particular through the medium of radio dramas. It does so by working with a rich archive of radio drama texts in isiZulu that has been largely ignored – by critics but not by consumers – as a key site of cultural production from the period of 'high' apartheid to the present. Finally, it moves its focus to the post-1994 era and sets the dramas within a wider arc of analysis, and in the broader context of a highly successful FM station, still part of the SABC but functioning in the democratic era. Alongside this, I refer to records from the radio archives of the BBC and the London-based Transcription Centre in order to include the two exiled writers, Modisane and Nkosi, within the arc of radio as a shared art form and a shared 'home'. I set this within the entangled transnational working of a segment of South African radio that was a product of the local, of nationalisms, of ideologies of race, and of imperial circulations that drew it always more deeply into the global world.

Radio Soundings: South Africa and the Black Modern is divided into three sections – 'Sound and "migration"', 'Distance and intimacy', and 'Drama, language, and daily life' – followed by a conclusion.

Structure of the book

Sound and 'migration'

K. E. Masinga has been viewed as hero, sell-out, unreliable trickster, dependable ally, brilliant translator across languages and cultures, and broadcaster par excellence. Chapter 1 interprets him as one of the architects of black radio in South Africa, driven by a passion for the new medium and the Zulu language. It describes the 'migrant' shift to the new medium in the face of political arrogance and the uncertainty of a World War, and then turns to Masinga's building of a new domain of radio with brilliant colleagues and new sounds and genres – all this achieved despite the power and cunning of the figures at the top whose aims were very different from his own.

Chapter 2 turns to the radio worlds of Alexius Buthelezi. Versatile, dedicated, and moving between the scripts of spoken radio drama and sung forms that verged on mini-operas, Buthelezi devoted 17 years to the new station that was designed to stunt rather than encourage any kind of renaissance of black talent. His contribution showed the irony of what had been intended and he contributed to a mini-renaissance of black culture that went unnoticed. He combined the skills of a choirmaster, author, and creator of new radio sound when he played a key role in making a drama working with an isiZulu translation of the Afrikaans epic poem *Raka* by N. P. van Wyk Louw. He then turned to a musical drama of domesticity and witchcraft, and followed this almost immediately with a mythical music drama about a young princess, *uNokhwezi* (Morning Star). This was entered by the SABC in the international Prix Italia competition in 1965, won an honorary mention, and thus moved into a transnational radio circuit well beyond the remit of the SABC. In doing so, it became part of a different South African national imagining. Alongside this, serial dramas about women in the city grew apace, gathering audiences and skilled performers, playing with language, and shaping ideas of the modern in the everyday.

Distance and intimacy

This section of the book explores radio 'intimacies of exile' through its focus on two South African writers, Bloke Modisane and Lewis Nkosi, whose work in radio is hardly known. Both wrote drama for the BBC. Bloke Modisane's small but important radio oeuvre was produced from 1970 to 1987, and, although he was completely outside the sphere of Zulu dramas that were running concurrently within South Africa, there are fascinating resonances as well as marked contrasts. Distance for Modisane meant involvement with the medium of radio for most of his life in exile, at first in Britain and then in Germany. Drawing largely on the records of the BBC, both written and sound archives, Chapter 3 extends our knowledge of the writer far beyond his *Drum* magazine years and his early autobiography *Blame Me on History* (1986 [1963]). It places him as one of the creative outsiders writing for the BBC at a time when drama on Radio 4 flourished under the innovative direction of Martin Esslin, and when radio as a medium was struggling to redefine its role in British culture and public life. His five BBC plays – *The Quarter Million Boys*, *Sitting Duck*, *Leave Well Alone*, *Freedom of the Vices* and the posthumously aired *The Somebody* – each show Modisane engaging with his distant 'home' as a diasporic, ironically nationalist figure, always in search of a community.

Chapter 4 focuses on Lewis Nkosi's varied links with the medium of radio beyond South Africa's borders. As Britain's African empire unravelled in the early 1960s, Nkosi played a key role in the dissemination of African and African-American ideas on culture and politics from his radio base at the Transcription Centre in London. Woven into this picture of the intense exchanges of ideas on Africa's future and past and the diasporic links to both the Caribbean and 'Negro' Americans are accounts of Nkosi's and Modisane's American connections, and, in particular, the influence of Langston Hughes and James Baldwin on the two men. Nkosi's BBC work, his play *We Can't All Be Martin Luther King* (1971), and the earlier short story 'The Trial' (1969) (see Gunner 2005) show that both Nkosi and Modisane, dynamic figures of exile, located themselves in a web of imperial, postcolonial, and transatlantic currents at a time of great flux and energy in world affairs. They did so in part through the access to new audiences and fluid cultural models that radio allowed. London itself as a city of changing racial profiles and restless cultures – and as home of the BBC – is part of the chapter's narrative.

Drama, language, and daily life

By the early 1970s, fear of 'the enemy to the north' stalked the corridors of the SABC. In spite of surveillance, the body of skilled writers and producer/actors was adored and followed closely in the 1960s, and by the early 1970s they had produced a thriving new genre that gripped the popular imagination. The audiences for Zulu serial dramas grew dramatically and their listeners were loyal but discerning and critical. Chapter 5 argues that the serial dramas became a key modernising medium at a time of embattled attempts at social control by the Nationalist government. Through a close analysis of three plays – *uBhekifa* (Look After Your Inheritance [a man's name]) in 1970, *Ubongilinda Mzikayifani* (You Must Wait for Me, Mzikayifani) in 1974, and *Abangane Ababi* (Bad Friends) in 1979 – the chapter highlights how the language of the plays shows the ways in which isiZulu itself handled new social pressures and transformations and how the plays 'talked' people through the tough but frequently dynamic years of the decade and made possible new kinds of 'addressivities' and new linguistic strengths.

One of the most powerful radio dramas in people's memory is the six-month-long serial *Yiz' Uvalo, Umanqob' Isibindi* (In Spite of Fear, the Victor is Courage). Chapter 6 asks how a serial that had 100 episodes and included some of the most loved actors and actresses on the airwaves achieved its status as compelling listening in the last months of 1986 and into the first half of 1987. It argues that the serial, by Morris V. Bhengu, tapped into the structure of feeling of the times, when civil war gripped parts of Natal and when the apartheid era was beset on every side. Moreover, in a decade when the field of talented performers and writers produced a wealth of dramatic fare, the combination of forces caught in this single long serial – of magic, modernity, and the sheer soundscapes of the episodes – meant that it became a classic of the era.

Chapter 7 explores the shift of focus in the radio dramas of the post-1990 era, often referred to in isiZulu as '*uhlel' olusha*' (literally 'the new arrangement'). The plays all feed into an understanding of how one lives in a nation just beginning to make a new history, yet with a past and present full of contradictions. The plays of the 1990s, within a popular genre more than three decades old, are full of violence and uncertainty. The rural space is marked with contradictions – nostalgia, humour, and bleak criminality are all present. The restless tropes of speeding taxis, consumerism, and modern marriage feature constantly. Yet, for listeners, what still seems to matter is not so much the ending as the process, the journey of the drama itself. In this decade, one of tumultuous change, optimism and despair are equally present. The dramas, which are fighting for listeners with popular television soaps as well as directed drama with specific messages, still find room for exploring ways of belonging and make space for the imagination of a multi-layered present.

This gives the reader an insight into the station from a vantage point outside the radio dramas. It shows how, in the second decade of the new millennium, the station has maintained a mission to create and serve a community of language speakers who stretch beyond first-language users and which covers a range of classes, of those inside and outside easily defined publics. It shows how the station itself, with one of Africa's largest listenerships, fashions its identity and speaks to its current vast audience, which includes those on the margins of South Africa's not so brave new world.

In Chapter 8, I ask how a radio station that had taken its listeners through so much in the apartheid years could continue to build a centre, or a series of centres, in a vastly different historical moment. I suggest that, in the new democratic era, the intimacies largely contained previously within the dramas became part of people's new public selves. They were voiced by listeners on phone-ins to Ukhozi FM, and in particular to the Vuka Mzansi Breakfast Show, which I discuss in this chapter. Also, the endlessly displayed selves in the domestic dramas in which the family was constantly tested and explored could now speak as individuals and tell parts of their own stories on air. In a sense, then, to draw from Laura Kunreuther (2014), they became 'voicing subjects' within a wider democratic space made possible by radio. This does not mean that the space of the domestic was lost within radio dramas themselves. The shifting site of marriage as a place of tested loyalties and conflicted ambitions continues to be explored in the final drama discussed in this study.

Part I

Sound and 'migration'

1 K. E. Masinga, Zulu Radio, and the politics of 'migrant' orality

'*Mkhulule*!!' [Get rid of her!!] K. E. Masinga's standard comment on the extremely popular phone-in programme that ran every Thursday evening at 8.30 p.m., *Ezinkulu* [Problems of the heart] on Radio Zulu.

The sounds of war: 1879 and after

This chapter explores ways of understanding the kinds of transformations and 'migrations' that occur when a language moves to a new medium, in this case radio.[1] I set out what particular tensions and plays of power operate when this occurs in an era when colonial, imperial, and, later, state power structures set languages in hierarchies of value. Through a close study of particular key moments in the history of the medium in South Africa, I discuss how radio in Zulu was able to adapt social knowledge and forms of performative power, creating a new 'vernacular' in a situation of unequal relations where silence and dancing to 'his master's voice' might have seemed an easy option. And through the case of a particular key radio 'migrant', King Edward Masinga, I ask if 'dancing to his master's voice' across languages and translations of culture can sometimes also be seen as a dance of power. What was in fact made through the new mediatisation of the black voice was a new set of publics within a wider public sphere, and new subjectivities (Kunreuther 2014).

One of the points made, often in the more obscure accounts of the Anglo-Zulu War of 1879–80 – obscure because the writers themselves were subaltern figures (for instance, Emery 1879) – relates to the 'noise' made by the Zulu usually before battles. Often the accounts were in the diaries of young Welsh soldiers, their minds 'edited' well before they reached the battlefield to know that they were fighting the black pagan

[1] An earlier version of this chapter was presented at the conference on 'Commodities and migration: things out of place' at the Department of English, New York University (8–10 December 2011). My thanks for the many useful comments from participants on that occasion. My particular thanks to Jo Valentino of NYU for his insightful comments.

subjects of the British Empire.[2] Sometimes there is a mention of the sound of song coming across the darkness or the moonlit spaces that separated the two armies. These sounds are described as unsettling, strange, eerie, forlorn even. Such songs or chants, which were most likely the *amahubo empi* (war anthems), would have been used along with the *izibongo* (praise poems) of former kings as inspiration for the men about to go into battle against an imperial power whose leaders had forced war upon them. Thus, what was understood as 'eerie and unsettling' by one side was a source of deep inspiration and wisdom for the other. As Fisher remarks (of indigenous Australian radio voices), to 'unsettle' can also be a means of affirmation (Fisher 2016: 14). For its makers and performers in the battles of 1879 through to 1880, the songs and chants were a mark of how one was in the world. The oral texts, because one can think of a song or chanted poetry as a 'text' (Barber 2007), travelled aurally and were an important part of anchoring the individual. But they were seen by imperial and colonial officials as part of the native exotic – and even, on occasions such as this, the threatening exotic. Certainly, the wisdom of the texts was seen as existing beyond and outside imperial discursive networks. Thus a binary was set up, and the premodern 'other' set in the imagination of the listener (Gautier 2014).

After the colonisation of Zululand and the breaking up of the Zulu kingdom in 1880, there was increasing migration to towns as men went in search of employment to pay the punitive taxes imposed by the imperial power. Nevertheless, the expressive forms that anchored people's understandings of who they were, such as praise poetry, continued to operate. As a way of shaping and stating one's identity they were vehicles for the construction of a social reality in which one had a place and an understanding of who one was. Praise poems, which were made for ordinary men and women as well as for Zulu royalty, constructed the personal and were composed by oneself and one's age mates. At times, praise poems encapsulated the fleeting sight of a dramatically altering social reality. They were part of the process of identity-making within fluid and even conflicted social systems. For instance, the life of a foreign black soldier in World War One could be drawn in a few telling poetic utterances. The praise poems from the Hlabisa chiefly house in northern Zululand capture such moments: the praises hold in their brief but dense and allusive images the huge disparities in how the owner of the *izibongo* is treated at home and when he is a non-weapon-carrying combatant for the Allied Forces in the 1914–18 World War:

[2] My thanks to Sue Krige for this comment on the 'edited' mentalities of the young Welsh soldiers and their diaries, one of which I have referenced.

The great ones of the Zulu set out and headed for England
When they arrived in England they became little 'nuisances' (Gunner and Gwala 1991: 125–6)

At approximately the same time, when Native Commissioners were established in districts of Zululand to dispense authority and the new power of the South African state, the strong awareness of these indigenous texts that validated social reality was turned by some Native Affairs officials into carefully documented accounts of the sonic and expressive world in which they found themselves. This was a social and political domain of sound that coexisted with the forms of paper bureaucracy that Native Commissioners had to administer and which their Zulu subjects had to negotiate and make part of a new vernacular. E. W. Grant was one such Native Commissioner who recorded on an early phonograph machine two elderly Zulu praise poets and published his account with the texts and historical background in the journal *Bantu Studies*. Calling his article simply 'The *izibongo* of the Zulu chiefs' (Grant 1929), Grant put on record the links between poetic form and rhetoric and political authority and so gave a glimpse of the social value placed on the utterances and their place as one of the many 'travelling texts' that populated the bleak and far-flung hills of the district in northern Zululand over which he presided. Clearly, Grant battled with how best to understand this; he may even have hoped to gain kudos among his peers from his 'native knowledge'. He published the poetry and the accounts of the elderly poets in a journal that was itself ambiguous in its attempts to engage with, comprehend, and publicise African social forms together with language, art, poetry, and narrative, all of which dealt with a different symbolic order. The journal, published by the University of the Witwatersrand from 1921, established a niche for some speculative and path-breaking studies within a wider context in South Africa in which hierarchies of race and knowledge were beginning to become normative.

The Grant article, like others in the early issues of *Bantu Studies*, unsettled rather than confirmed the view of a solid imperial world of knowledge and racial superiority. It did so somewhat obliquely through the ways in which it pointed to the presence of the sonic as a medium of a social world and ethical values, of travelling sound as text and its operation as organiser of social knowledge and memory as well as a maker of the everyday. The power of the form – its stately poetic quality, its commentary on history and character – was present, and the echoes it contained of the epic and minstrel traditions in Western verbal art were recognisable. It was thus presented by Grant as an item of arcane dignity but ambiguous in terms of value. Were the long stately poems collectors' items reminiscent of a lost precolonial order? Or were they something else? Were they perhaps a gesture to an alternative knowledge structure and sound system? One that, in the days of late empire, knocked against bureaucratic missives of the Department of Native Affairs, yet for much of the time had its own existence where it was part of a technology of operational networks circulating along other lines?

The same ambiguity in understanding and placing items from the older sonic and social orders was present in the modernising African communities of the second half of the nineteenth century that were often attached to mission centres such as the American Zulu Mission at Inanda, close to the town of Durban (Hughes 2011). The *izibongo* and other forms of oratory, such as clan anthems (*amahubo*), dance chants (*izingoma*), and folk narratives (*izinganekwane*), to name but a few, were frequently discursive mediums and fluid genres that allowed a grappling with modernity but did not jettison understandings built over generations. What, in the latter decades of the nineteenth and early twentieth centuries, was often referred to as the *kholwa* community – the mission-educated African men and women, personified by men such as John Langalibalele Dube – also had to negotiate a way of taking on the value of older technologies of the sonic. John Dube, who was to become the first president of the ANC in 1912, was also deeply influenced by his interactions with African-American intellectuals such as Booker T. Washington. Partly through his pride in the African race imbibed during his stay in North America from figures such as the Washington-based African-American intellectual and distinguished journalist John Edward Bruce (Hughes 2011: 72–3), Dube was able to confront the increasingly sharp racism of white Natal settler leaders on his return home. He could also fold into this race pride a place for the discursive forms I refer to above, which were often bundled together as 'folklore', a term that in official parlance safely distanced them from the modern. Such forms in the cultural archive depended in the first instance on orality and can be understood as social networks of sound and a social construct of technology. Dube thus endorsed, even if with some ambivalence, the value of the forms of sound while working towards the founding of a modern nation (ibid.: 261). Sound texts, one could say, were both within and without the normative understanding of 'the modern' in Dube's practice. As a social commodity of uncertain but definite value, they were perhaps something of a wild card. They can perhaps also be seen as migrant texts, travelling texts, capable of re-emplacement, adaptations, and new vernacularisations.

A new medium: radio, a new migrant

Radio and its soundwaves, in the South African context, found their way quite slowly into this unsettled world of the early decades of the twentieth century. In the new medium of radio, the travelling texts of orality were to find a place but the journey was not a simple or an easy one. The routes by which radio found its niche in South Africa were linked to the circuits of empire. In its early manifestations in South Africa and its broadcasts from makeshift technical rooms, first of all in Cape Town and Durban in the 1920s, with the first Durban broadcast coming from the City Hall in June 1924, radio sent out shards of political knowledge and popular culture from the British heart of empire to English-speaking settlers and to migrants from Europe for whom South African English became the new lingua franca (Rosenthal 1974).

Afrikaans, which was beginning to shape itself as an evolving modern language tied to the new 'volk', was largely silent on these early airwaves, as were the African languages of the region. For a brief period in the mid-1920s an African variety programme featuring the 'Zulu Versatile Company' had a slot in the Durban radio airtime, and Eric Rosenthal notes that it was also popular with many English-speaking listeners (1974: 97). However, the first programmes aimed 'officially' at Zulu-speaking listeners did not come on the air until December 1941. As I mention in the Introduction, they seem only to have happened at all because of the ruptures of war and possible anxieties about a restive 'native' population at a time of crisis. Yet, in the way the narrative has been handed down, almost as an urban folk legend itself, it was the initiative of a single individual performing his way into the portals of the SABC Durban studios in Aliwal Street that began the Zulu airwaves sound. King Edward Masinga, a young headteacher who trained at Dube's Ohlange Institute and obtained his teacher's qualification from Adams College in around 1924,[3] desperate for the war news to be broadcast to Zulu listeners in their own language, walked into the Durban SABC studios and asked to be employed for this purpose as the first Zulu announcer. He met with a refusal, but instead of walking out sadly – or angrily – he performed his exit within a particular genre. Drawing on a set of gestures and an array of semiotics recognised as part of the powerful native exotic, he walked out backwards from the presence of the man who had said 'No!' (as one would when leaving the presence of Zulu royalty). As he did so, he exclaimed, as one would for Zulu royalty, '*Bayede Wena weNdlovu*! [Hail, You of the Royal House!; literally, 'of the Elephant'] *Bayede*!' At which point – so the narrative goes – the official figure of authority (undoubtedly the journalist and ethnomusicologist Hugh Tracey[4]), changed his mind, and a deal was struck.[5]

[3] Masinga interview on SABC website.
[4] For a fascinating account of the much longer musical collaboration of Masinga and Tracey, for which Masinga appears to have received no credit, see Mhlambi (2008).
[5] 'Izihlabani Zomsakazo' uMnu K. E. Masinga', Acc. 65982, Cat. T (Z/94) 161, 1986, SABC Sound Archives, K. E. Masinga Road, Durban.

With this version of how isiZulu found a spot on the airwaves, the trope of the ambiguous, potentially powerful outsider, balanced by that of the trickster figure, came into play. Each remained operative ways of articulating, directing, and imagining African-language radio – or 'Radio Bantu' as it became officially known in 1960. Like a character in an epic or folk narrative, Masinga found an entry for himself and his 'people' into the magical but dangerous kingdom of radio through flattery, and by performatively recalling a set of gestures that were already quite anachronistic but were nevertheless still full of symbolic valence and affect, and linked to memories of old power and modern nationalism (La Hausse de Lalouvière 2000). Each trope – that of the powerful outsider and that of the trickster figure – can be seen as variously illuminating different aspects of how 'the native' took to the air in ways not entirely dictated by the 'master'. The breaching of the portals of the Durban SABC marked, in a way, the moment when the texts of orality migrated into the new aural medium and signalled the beginning of what we could call a new Zulu social order of the air. To put it another way, it was a kind of migration. It marked the beginning of a new ideology of voice, of voice as a different kind of expressive agency. It was perhaps unsettling to some as well as deeply satisfying to others. Kunreuther notes in relation to the entry of FM radio into Nepali life in the 1990s that, through voice, Nepalis came to understand themselves as 'voicing subjects within the limits and possibilities of a host of emerging publics' (2014: 9). Masinga's entry was half a century earlier and into a deeply undemocratic moment. Yet it was to create changes of a similar magnitude.

Masinga: the trickster and maker

The personality who dominated the new venture was K. E. Masinga himself. His first broadcast in December 1941 was a three-minute news clip on the progress of the war. These short news items in isiZulu became a regular feature of the service from Durban and gradually increased in length to five minutes, with other items such as 'sketches, dramas of folklore in Zulu, then dramas on crime does not pay and health programmes' all added by Masinga himself, although he must have been aided by Hubert Sishi, who joined the Durban SABC to work with Masinga in 1945.[6] Masinga clearly had ambitions for a Zulu radio service that extended far beyond such items of information, no matter how important these were in the context of South Africa's fortunes in the uncertain arena of World War Two. The question, for our interest in the

[6] Masinga interview on SABC English Service, SABC Sound Archives, *c.*1958.

Figure 1.1 The King! The broadcaster King Edward Masinga broadcasts on Durban SABC's Bantu programme, *c.*1956.
Source: *Drum* photographer, BAHA/AMO.

migrations of culture and the setting up of African-language radio, is how he tried to create a public media space for isiZulu in a new global medium. Moreover, as the Zulu language on radio, it was also to become an increasingly important national medium of communication and identity formation when the odds seemed stacked against him. How did Masinga manage it? His first real move in carving out a space for texts that were in one sense 'folk' yet were also embedded in the new medium occurred in about 1943 (the precise date cannot be confirmed), when, with Hugh Tracey, he produced the musical script of a folk tale, *Chief above and Chief below*, with 'original songs by the author in traditional idiom' (Tracey and Masinga: 1945; Coplan 1985: 160; Gunner 2000a: 234; see Mhlambi 2008; 2015). The recasting of an oral story in the new medium was hugely popular, particularly with migrant hostel dwellers in Durban who listened not through wireless sets but by means of rediffusion, which transmitted the broadcast items via cable to a central distribution point. This, then, must have constituted a kind of public–private listening space where groups of hostel dwellers would listen

Figure 1.2 K. E. Masinga, senior announcer of the SABC's Zulu programme, with his colleague, Hubert Sishi, discuss the programme for the November 1957 African Arts Festival in Durban, October 1957. They were chairman and secretary of the festival.
Source: Ranjith Kally, BAHA/AMO.

together, respond to the radio items, and discuss them both individually and collectively. Masinga, eager for 'a fully blown Zulu broadcasting service for the black community',[7] did not impose his ideas on listeners, but rather, with his soon-acquired colleagues Guybon Mpanza and Hubert Sishi, sought the views of those at whom the broadcasts were aimed. This meant moving around in workers' hostels and in factories, gathering listeners' views and needs (Couzens 1985), and turning their finds back into use in the snippets of Zulu-language airtime allowed them.

The migration of the established oral form to a new sound where voice became mediatised was clearly intrinsic to Masinga's vision for a classical modern Zulu culture that was part of modern technology and, because of its oral base, was particularly well suited to radio. What Masinga may not have always emphasised as he carried out his balancing acts with his austere and undoubtedly powerful SABC 'masters' was that popular culture and popular music – both Zulu forms such as early *maskanda*, which he would sing himself with banjo accompaniment, and singers

[7] Interview with Masonto Buthelezi, 1999.

such as Frank Sinatra – were part of the new programmes. These helped create a sense of popular culture and the modernising black world – a world in which, at times, the urban and the rural fed off and into each other.

In the course of his long career as a multiskilled broadcaster working with the Zulu language, Masinga seems to have performed many careful balancing acts. As he found throughout the whole of his career, which ended formally in 1971 but went on informally much longer, this was imperative when constantly working with powerful superiors in a racial hierarchy of radio where voice was always black but the controllers, both technical and in terms of authority, were always white.[8] This quite chilling and seemingly unbreachable hierarchy of race can be seen, for instance, in the congratulatory reports and pictures in the *SABC Bulletin* of 5 February 1962, which reports on the birth of the 'Commercial service of Radio Bantu'. Two pictures show black announcers in front of a microphone. In one, Bhekisisa Kunene (later well known in radio drama), is at the desk for the 'first day-long Zulu Service'. The other photograph features two announcers, Petros Tumane and Thabo Chaka, who were 'responsible for the first South-Sotho broadcasts from Johannesburg'. Yet the real power appears to lie within the three pictures of white officials in suits and ties, in two cases holding cups of tea and chatting to each other, and in the third 'watching the first broadcast'.[9] The question remains: did they know they were 'sewing dragons' teeth'? Clearly not.

Masinga's guile and determination to pass through forms of control to create the free space of imagination that a musical such as *Chief above and Chief below* and other subsequent items brought to the airwaves were impressive in their own right. Yet they seem even more impressive when set alongside the crude official attempts to use the airwaves, with language regarded as a simple and singular sledge-hammer medium for information and communication alone, to be directed at a subject population. In this vein, the earliest archived recording of isiZulu featuring alongside English, Afrikaans, Sesotho, and isiXhosa is in the Johannesburg Sound Archives of the SABC in Auckland Park, dated September 1942.[10] The strange and strained register of the announcements, which are introduced and concluded by 'African' sounds (a clip

8 Louis van Rensberg, Radio Production Services manager (SABC Auckland Park), personal communication, 28 November 2011. My thanks too to Mr van Rensberg for explaining to me the basics of radio soundwaves and the FM system of radio waves, whereby soundwaves move 'in a straight line'.

9 *SABC Radio Bulletin*, 5 February 1962, p. 6, 7. Courtesy SABC Library, Auckland Park, Johannesburg.

10 'First broadcast to natives', 1/63 (42), 29 September 1942, SABC Sound Archives, Auckland Park, Johannesburg.

of men's voices in antiphonal song with rhythmic handclaps, a flute melody, a lion's roar), was perhaps an attempt at African-sounding rhetoric, but a white voice of authority delivers a wartime message about the extension of radio broadcasts to 'the Abantu [people] of our land'. Extracts from the English announcement (which followed that in Afrikaans) are as follows:

> There will be broadcasts for Natives from today. Hear this word which makes known in the land that by the kind-heartedness of our country's wireless system a boon is being granted to the Natives of our land, namely the Abantu of the Union of South Africa ... We begin this morning with an explanatory statement by the Department of Native Affairs ... Because of the generosity of the SABC from today 30 minutes on Tuesday, Thursday and Saturday will be broadcast at 9.30 in the morning by the SABC for the Abantu.[11]

It must have been clear to the astute black listener that only some kind of perceived national crisis had occasioned this largesse. Harcourt Collett, the announcer for the English segment, with his high-pitched, strangulated BBC-like voice, so similar in its speech register and authoritative tone to other African British colonial radio services (Bloom 2014), continued to inform 'the Natives':

> Yes indeed there are still loudspeakers in the big cities in our country. They will go on, set there by the Native Affairs Department [NAD], but the different programmes on the air will start today. There will be nothing on holidays or on Sundays or only if considered to be necessary for the country as a whole as an emergency.[12]

In a sense, this was shutting (or perhaps opening) the stable door after the horse had bolted. Masinga and his new colleagues were already busy with war news and other items for Zulu listeners of the SABC, working from the Durban studios, and the elaborate Masinga–Tracey radio recasting of the story *Chief above and Chief below* would soon be on the air. As I point out in the Introduction, the deeper unease about controlling a subject people in a time of crisis was indeed behind not only the September 1942 announcement but also the December 1941 Masinga breakthrough. It appears that Hugh Tracey, picking up on an early example of 'radio trottoir' (Ellis 1989), had persuaded his superiors at the Durban SABC that airtime in isiZulu was essential as a means of ensuring that the rumours in the black community of an imminent Nazi

[11] Ibid.

[12] Ibid. My thanks to Peter Raseroka, 'Uncle Peter', and Rietha Buys of the SABC Johannesburg Sound Archives in Auckland Park for their generous help and for making this recording available to me in November 2011, as well as those in which K. E. Masinga features, which I mention elsewhere in the chapter.

victory, which began to surface in South Africa during the Russian campaign, were quashed (Coplan 1985; Couzens 1985).[13]

Wireless sets as commodity, sound as commodity

Although, in the early 1940s, rediffusion was most common among black listeners and *amawayilense* (wirelesses) were unknown outside the small African elite, all this was soon to change. The writer and journalist Herbert Dhlomo reported in the Zulu-language weekly *Ilanga lase Natal* in 1948 that 'thousands of Africans have receiving sets' (Couzens 1985: 237). The heavy items of furniture that constituted wireless sets in those days would probably not have been found in many black homes. It was sound itself rather than the sets that was the important commodity. The sound, which I argue had in some way migrated to radio texts, was itself a heady commodity comprising not only transformed oral narratives, histories of the Zulu past as drama (in this both Masinga and Sishi excelled; see Sishi 2000), and varieties of song and poetry, but entirely new sound objects too. Masinga and his early radio colleagues were migrants and adventurers rather than antiquarians.

There was also the issue of translation, and cultural translation. There is virtually no archival trace (as far as I am aware) of the nine Shakespeare plays that Masinga translated and that were put out in the Zulu programme airtime, probably over the English service of the SABC. The Shakespeare plays seem to have been on the air in the early 1950s, as Masinga was elected to the International Mark Twain Society in 1953 and in the same year his translations of *King Lear* and *The Tempest* were serialised on the SABC (Coplan 1985: 161). Masinga himself refers to his translations and renderings of *The Merchant of Venice*, *Romeo and Juliet*, *A Comedy of Errors*, and *Macbeth* in his interview on the English Service in the 1950s.[14] Yet all that remains is a spoken four-minute extract, in English and isiZulu, of Mark Anthony's famous 'Friends, Romans, countrymen' funeral oration from *Julius Caesar*. Masinga recited the isiZulu with great dramatic vigour – '*Zihlobo, bakwethu, maRomani* . . .' [Friends, Romans, countrymen . . .] – the actor Ian Calder recited the English far more calmly. The snippet was archived in both the English and Zulu service and aired – perhaps on both programmes – on 18 March 1955.[15] Was this, too, an instance of a kind of cultural translation, where the political weight of something approaching Zulu praise poetry could be

13 Interview with Masonto Buthelezi, 1999.

14 'Malcolm Fleming and K. E. Masinga', Acc. 201420, E86/97, 24 January 1958, SABC Sound Archives, Auckland Park, Johannesburg.

15 '*Zashonaphi Izihlabani* [Where have all the brave ones gone?]', Cat. T (Z/94) 23, SABC Sound Archives, K. E. Masinga Road, Durban; 'Julius Caesar', E86/86, 1955, SABC Sound Archives, Auckland Park, Johannesburg.

Figure 1.3 'Enjoy both with an Eveready Spacegram', *Ilanga* newspaper advertisement, January 1960.

Figure 1.4 '*Yonga imali*!: Something to save for, something to treasure', advertisement from Natal Radio and Furnishers, Durban, in *Ilanga* newspaper, January 1960.

felt in the revolutionary speech of a Roman politician, itself a 'translation', of a kind, by a sixteenth-century English dramatist?

If we can think of sound, in the particular instance of radio, at a particular time, and within a certain frame, as a commodity, we can perhaps also understand it as a contested commodity. Saul Dubow has noted how the ideas of racial segregation that marked the Smuts (United Party) years of government in South Africa (which ended with the 1948 election) were characterised by 'ambiguity and ideological flexibility'. This differed markedly, Dubow argues, from the National Party ideology of apartheid with its 'unremitting zeal and logic' (1989: 178). Yet both Deborah Posel and Dubow have argued that there was no 'grand plan' of apartheid. That widely held idea is a myth. Posel suggests that a close look at the workings of the state reveals a series of struggles, with revisions and adaptations. There were uncertainties, conflicts, factions, and deviations. She notes that, in the early 1950s, the threat of African mass resistance was still present, with serious gaps exposed 'in the state's control over the urban proletariat' (Posel 1991: 5, 6; Dubow 2014: 293). Radio played a role in this.

Expansion, control, and 'straight lines'

> The problem is that the moment we give the Native a radio and encourage him to listen, we place in his hands a weapon which may harm the Native himself.[16]

By the time the Hon. Minister Herzog made this remark in parliament in June 1959, the truth on the ground – as opposition members of parliament repeatedly pointed out – was that many thousands of African listeners already had licences for radios. 'Natives' were using the English and Afrikaans services of the SABC and Springbok Radio, the commercial station available, and many were also listening to the Zulu and other African-language programmes available in the time allotted. In addition, there was a vibrant and very popular rediffusion service that was being used in Orlando and Jabavu townships on the West Rand, where lines went straight to people's homes. It was run by the Orlando Rediffusion Service Ltd with the agreement of the SABC[17] and provided listeners with a wide range of programmes, including some from outside the SABC. There was popular pressure for the service to be expanded to neighbouring townships and the Orlando Rediffusion Service was keen to oblige.

[16] Dr J. B. M. Herzog, Minister of Posts and Telegraphs, South African House of Assembly, 3 June 1959, Hansard 7221.

[17] State Archives/NAD/Box 2793, 16 March 1959.

The appetite for radio and the commodity of sound was huge. A company called Titan Enterprises and Exploration, based in Boksburg on the East Rand, was keen to sell radios 'in Bantu Areas' and in early 1960 applied to the NAD for permission to do this. The response to this request was astonishing. Not only was there a refusal, but a letter was sent to Chief Bantu Affairs Commissioners in Bloemfontein, Pietersburg, Pietermaritzburg, Potchefstroom, Umtata, and King William's Town informing them of the situation. In other words, as if a severe epidemic was about to hit the land, the controllers of 'Native Affairs' in far-flung corners of the country – the centre, north, south, and east – were prepared, warned, and informed of the decision that this was *not* to be allowed. Radio sets were *not* to be sold to 'the Bantu' who lived in areas – urban or rural – that were designated as being under the control of Native Commissioners. Titan Enterprises was informed, politely, of the decision and no reason was given.

The smooth and unctuous words of the minister in parliament in 1959, quoted at the start of this section, contained an assumption of the truth of rigid racial hierarchies and the ease of control. Yet the statement hid, perhaps not very well, a profound sense of panic about radio. The National Party government had laid out spatial control of segregation through post-1948 legislation as far as it could, but radio seemed to need control of the atmosphere, the air itself. How was this to be done? Radio, moreover, was a medium that defied even spatial boundaries through the freedom of imagination, conversation, ideas, and knowledge which it enabled, something that I write about in more detail in Chapter 5. The National Party plan was to obtain an amendment of the 1936 Broadcasting Act (inspired by Lord Reith of the BBC) and to institute a Bantu Programme Control Board (BPCB) that would be separate from the rest of the SABC. This would control programmes for the 'Bantu'. In this way, the National Party, and principally the new prime minister, Dr Verwoerd, hoped to control the sound and voice, and everything else, that went over the air to the 'Bantu'. New transmitters were to be built that would allow FM transistor radios to be used – transmitters that would send out signals at very high frequency so that the reception was excellent. These signals would reach listeners in even the remote rural areas. The new FM transistor radios would receive the radio waves that came in 'straight lines' (a phrase the minister used about the new system) but would not pick up other radio signals. It was a wonderful and devious plan to control the airwaves. The powerful transmitters were put in place in 1962, and in some ways they were very effective. In others, though, the plan was as futile as it was vicious. It could work in rural areas and give excellent reception for a limited range, but in urban areas it was absolutely unworkable because of the proliferation of signals that listeners could pick up.

The amendment to the Broadcasting Act was passed in the face of intense and passionate – and also very well informed – opposition in parliament. The attempt to control the commodity of sound, and to optimise already far-reaching levels of surveillance, went into national operation. In June 1960 a new radio station named Radio Bantu came into being which would broadcast in Zulu and in some of the other main African languages, in the first instance Sesotho and isiXhosa. Radio was to become indeed a 'weapon', but one to be used by the government.

Again, the metaphor of the bolting horse and the stable door closed too late comes to mind. Was all the above, the elaborate plan with its longing for 'straight lines' rather than crooked airwaves, too late, as the powerful radio stations of independent Africa – Ghana, Radio Cairo, and others – beamed their signals across borders? Was it also too late as, by this time, the Zulu service that had begun in the early 1940s (with other African-language services – South Sotho, isiXhosa, SiNdebele, Xitsonga, and Setswana – coming later) had established its own voice and body of informed listeners? These services were already working with new radio genres, translating at a number of levels, mixing and remixing cultural forms for a new community of the air. Was it a case of too much too late?

The answer has to be yes and no. Surveillance, long in operation, continued and was intensified as the positions of white controllers became more formalised and the Board for Radio Bantu was set in place. Programmes were monitored, and so were the black staff. And yet, in some ways, perhaps it *was* too late. A migration had taken place. New voices of cultural authority were on the air. The ambiguous domain of 'folk' art had been remade and remixed alongside new forms, and the trickster figure of King Edward Masinga, 'migrant' and maker, had survived. Surveillance had stalked him, too. In 1957, when the United States government invited him to be a participant in the Foreign Leaders Programme of the International Educational Exchange Service, he was given a passport but only after police clearance. Moreover, he had to put down a deposit of a considerable sum of money that was only returned once he was home and had handed back his passport, when he became once more merely a 'Bantu' subject and not, officially, a South African citizen.

While in America he lectured at a number of universities: 'Hertford University, Howard, Boston, and some universities also in Washington and Indiana,' he explained to his English-service SABC interviewer on his return in January 1958.[18] Masinga, never one to favour modesty,

[18] 'Malcolm Fleming and K. E. M. Masinga', Acc. 201420, E86/97, 24 January 1958, SABC Sound Archives, Auckland Park, Johannesburg; 'Malcolm Fleming and K. E. M. Masinga', Acc. 57113, Cat. T (Z/94) 23, 21 January 1958, SABC Sound Archives, K. E. Masinga Road, Durban.

explained in reply to the carefully phrased, and loaded, question, 'Did you have time to travel?':

> I think I got all I wanted ... I had all the travel facilities laid on for me. I travelled from coast to coast of the United States. Most of those journeys ... were to TV studios, broadcasting and mostly universities. I had no time for sound broadcasting. I was rather interested in television itself ... at first all those lights were really a little bit dazzling but I soon got used to them.[19]

Masinga informed his interviewer that he had lectured on African folk songs and music and on African social anthropology 'that was of high interest in America', and that he also conducted research in the southern states: 'Alabama, Virginia, Arizona [sic] where the Negro lives in the rural areas'. He complained about accent and obliquely touched on deeper questions of race, identity, and belonging in America:

> One handicap was that the Negroes have got a certain growling and nasalising dialect ... a drawl which I do not understand readily ... [and] the Northern Americans I am afraid find it very difficult also to understand one another.[20]

Although not able to control his travel in North America, the South African government, as mentioned above, made sure that Masinga returned his passport when back in the country. But with the visit Masinga became yet another of the many South Africans – including John Dube – who in the nineteenth and twentieth centuries had made the transatlantic journey and shared knowledge of social and cultural forms across the gaps of history and the attempts at domination. Unlike his contemporaries, Bloke Modisane and Lewis Nkosi, neither of whom was allowed a passport, he seemed to be able to move beyond the closed borders of the country with a kind of untouchable bravado. Again, it was perhaps a dance, a tricky performance, but inspired and courageous. Modisane and Nkosi, whose own radio lives were made outside the country and who are the subjects of Chapters 3 and 4 respectively, were faced with different challenges connected with race, power, and culture. These they handled with their own combinations of ingenuity and panache. For each of them, too, as for Masinga, voice performed many functions in which agency was a central feature.

Language, power and 'the King'

On 31 May 1961, South Africa became a republic and thus threw off the last formal vestiges of the British Empire. Radio Bantu and the African-language services, including Radio Zulu, continued. Masinga stayed in place, subordinate but in some ways always 'kingly'. To the

[19] Ibid. [20] Ibid.

end, he saw the space between languages and cultures as one for negotiation, power, and play. When, in 1970, he was awarded a 'prize' by the 'Nguni Services', he masterfully turned the tables on his white 'masters' as he 'thanked' them for his prize. He did so by using the poetic language of *izibongo* to name each of his superiors, and he thus captured them in the image of the 'other', seemingly subaltern, language. In his acceptance speech, given almost 30 years after his first bold entrance through the portals of the SABC in Durban, he moved between English and isiZulu. As he switched between languages, a colleague translated for him – but he was the master. The harshest praise name of all was kept for a Mr Louw, a senior official of the station. Mr Louw's praise name of *Umthakathi* (Dangerous Wizard) is left untranslated in the English, and Masinga's comment in isiZulu is harsh and yet opaque and gives little ground:

> *Ngiyakubonga Mthakathi. Sonke siphikisana ukuvumelana.* [I thank you Mthakathi. We all have intense disagreements but in the end we manage to get along.]

2 Communities through the back door

The radio worlds of Alexius Buthelezi, 1961–78

I was at medical school and someone phoned me and said, 'Your father's house is on fire.' It was 1981. Thami Buthelezi[1]

He was 'one of the pioneers in radio broadcasting . . . [with] a lot of love for the Zulu language and its culture and all of that.' Koos Hadebe[2]

Remembering the past, making the present

In 1961, a year after the Sharpeville massacre of 22 March 1960, which left 67 protestors dead, teacher and choirmaster Alexius Buthelezi moved from Adams College, Amanzimtoti, Natal and took up a post as broadcaster in the Zulu-language section of Radio Bantu in Johannesburg. It was not an auspicious time. Prime Minister Hendrik Verwoerd showed his intention to make the SABC work on *his* terms by summarily firing Gideon Roos that same year. Director general since 1948, a nationalist but also a Rhodes scholar, Roos was a liberal, and a fervent admirer of the BBC and its Reithian policy of public broadcasting. Roos had insisted on referring to black men as 'Mr' on all SABC stations. This public affirmation of respect and assertion of equality across races so infuriated the prime minister that he fired him in his absence and Roos received the news while on a family holiday.[3]

All black employees of the SABC entered through the 'back door' in Troy Street, rather than through the front entrance around the corner in Commissioner Street.[4] So Alexius Buthelezi entered the SABC by the same route. One could say, though, that he did not come alone; he brought whole communities with him. A pivotal figure in the dynamic, conflicted space of the Zulu services of Radio Bantu, and later Radio Zulu, Buthelezi's versatility exploited the nature of radio as a key technology of modernity as well as the fluid boundaries of cultural forms.

[1] My thanks to Thami Buthelezi for the memories of his father and his life at the SABC shared in an interview in Scottsville, Pietermaritzburg, 26 June 2012.
[2] Interview with Koos Hadebe, November 2000.
[3] SAMRO documentary: 'SAMRO 50 years'. My thanks to SAMRO for access to this.
[4] Interview with Winnie Mahlangu, 2006.

Figure 2.1 Alexius Hezekiah Kusa Buthelezi, composer of music dramas for radio, producer, and actor – a key figure in Radio Bantu and Radio Zulu.
Source: Reprinted with the permission of Dr Thami Buthelezi.

If Buthelezi was caught in an imposed Verwoerdian frame of 'Zuluness', he was at the same time, along with his isiZulu-speaking radio colleagues, redefining and reshaping the term for a modern era. The ambiguities and complexities of Buthelezi's position as a creative but 'controlled' radio man of the 1960s and the 1970s form the basis of what I explore in this chapter.

Buthelezi's contribution over his 17 years with the SABC was rich and diverse. One strand of his work focused on musical drama and myth, a path that K. E. Masinga and Hugh Tracey had first marked out in the early1940s (see Chapter 1). An important aspect of Buthelezi's life as a broadcaster was his musical innovation as interpreter of the reach of sound in creating new forms with which his listeners could interact on terrain that was both familiar and also strange, and perhaps entrancing and enticing. Alexius Buthelezi formed the Radio Bantu Choir shortly after his arrival, and one of its activities was to perform in the radio music dramas with which Buthelezi was closely involved. First came *Raka* (1963), then, far more importantly, *uNobathakathi* (1963) and *uNokhwezi* (1964), with his last music drama being *uChakijane neSalukazi* (Chakijane and the Old Woman) in 1975.

Buthelezi was steeped in the rich *amakwaya* choral tradition, which grew from the teaching of music at the early mission schools. His own career before he joined the SABC was as teacher and choirmaster, first at the well-known Catholic Marianhill School in Durban, and then at the equally prestigious Adams College, south of Durban. By the time he took up his position at Radio Bantu he was widely respected as a conductor and adjudicator and had his licentiate in Voice from the Royal College of Music. Choral music-making with its roots partially in African archives was extremely popular in African schools. African teachers, having absorbed choral music from mission education, had in their turn introduced school choral competitions, which worked regionally and came together regularly as Eisteddfods at a national level. Buthelezi was a key figure in this world. He was, though, equally steeped in the songs and dances of his childhood district of Pomeroy, in Msinga, which had as their cultural referent the idioms of a distinctive Zulu musicality not easily captured in the melodies and harmonic structures of the *amakwaya*.[5]

Buthelezi's range across diverse musical idioms that cut across any idea of 'high' and 'unofficial' culture enabled him to imagine newness in a particular way. As a man from Pomeroy with his own praise name, *Zigi zendoda* (Man with a heavy step!), he crossed boundaries and social groups with ease. The family home in Diepkloof Zone 1, Soweto, was full of the music of choirs, both the choral kind and those of *isicathamiya* singers who would visit for a last practice before a competition and sometimes sing and prepare 'even in the bathroom'.[6] It was this ease with diversity that he put to use in making his music dramas for the newly available cultural space on Zulu-language radio and for the reach of his *Cothoza Mfana* programmes, begun in 1962 (Gunner 2014). Both his workplace at the SABC and his home in Diepkloof became key sites of his musical creativity in a very public way.

By the time *uNokhwezi* was produced in the mid-1960s, the intense jazz life of Sophiatown and other townships of the Rand was over; African musicians were hampered by the pass laws and permit restrictions on performances. Alfred Herbert's African Music Variety shows of the early 1960s had been and gone; the jazz musical *King Kong* had left for London (Coplan 1979: 159–60), and the recording companies were exploiting 'the talents and complete dependence of their musicians' (Coplan 1979: 162; Meintjes 2003). As director of the Radio Bantu Choir, and with the opportunity to work from Johannesburg on new Zulu programmes, Buthelezi had the chance to graft the choral sounds into the new form with which he wished to work. He had radio as his medium, and this may have given him composing and artistic freedoms that would otherwise have been unavailable to him. He could project his work to a diverse audience of black listeners across class and draw together previously

[5] Interview with Thulas Nkabinde, 2007. [6] Interview with Thami Buthelezi, 2012.

scattered audiences in shared time and a created spatial moment. Sound for Buthelezi worked in diverse ways – through musicality but also through voice as a dramatic broadcasting medium cutting across the rural and urban space, and often honing in on black urban life and popular culture. It was a confident sound-space, mirroring the variety of opinions and writings of black voices in papers such as *Bantu World* and Durban's *Ilanga laseNatal.*

Searching for the new: the music drama

Buthelezi's music dramas attempted to fuse the energies from different cultural performative sites into a single new genre through the medium of radio. He was shaping a different kind of appeal, creating a longing for a new form in a still growing public. It was, too, perhaps, a chance to find a national vision, a past that could be worked into a new present. This was a daunting task but it was a view of a modernity that had much in common with the earlier strivings of writers such as Petros Lamula (1924) and Herbert Dhlomo (1985), who some years earlier had searched for an African present that embraced aspects of the past yet avoided ethnographic essentialism. It was celebratory of 'being Zulu' and also of being a modern African – the 'New African' as espoused by Dhlomo (see below) and other intellectuals such as Walter Nhlapo, a columnist with a special brief to cover music for the Johannesburg-based *Bantu World* (see Coplan 1985; Thomas in Cole and Thomas 2008).

Much had already been achieved in terms of newness by 1961. The audience for Zulu-language radio was already firmly established (Chapter 1): radio personalities such as Dan Twala, Joshua Mlaba, and Mandla Sibiya had already etched their voices and their styles of presenting on their listeners' memories and tastes. Dan Twala brought with him his extensive earlier experience of drama from the Bantu People's Theatre and the lustre of having played a lead role in the political realist play *The Rude Criminal*, where Sesotho was used even more than English (Kruger 1999: 75). As the broadcasters and 'creators' on Zulu-language airtime – who were often also the actors, narrators, and scriptwriters of these early dramas – jostled in the crowds from Park Station, central Johannesburg, and entered the SABC building through the doors in Troy Street reserved for black employees, their listening communities came with them. Black and white bodies were to be kept out of intimate physical contact, or closeness. Sound, though, could not be as easily separated as bodies and matter, and the essence of radio was sound. Beyond and yet within the body, sound could shape affect and ethics and create an intersubjectivity that took the individual out of a solitary world into a shared one (Kunreuther 2014). It was thus an intensely creative shared space, a forum – *inkundla* in isiZulu, which was also the title of Jordan Ngubane's dynamic newspaper of the 1940s and 1950s. But it was not visible and therefore not easily policed.

The year 1963 saw the production on Radio Bantu of two music dramas with which Buthelezi was associated. In the music drama based on the Afrikaans epic poem, *Raka*, by N. P. van Wyk Louw, Buthelezi's creative role may have been limited to conducting the Radio Bantu Choir, which performed the songs. The music for the Zulu *Raka* was composed by the Johannesburg-based P. J. Simelane. The direction, and possibly the shape of the text, came from Trudie Pohl, daughter of Anna Neethling-Pohl, doyenne of Afrikaans radio and a staunch Afrikaner nationalist (Kruger 1999: 39). Mandla Sibiya is also listed as a producer for *Raka* and possibly took the part of narrator – the scanty records are unclear on this. *Raka* (1941), a long, epic poem that made its author famous, can be read as informed by the 'scientific racism' of its time; it is a text galvanised by the 'menace of mimicry' and simply 'a high-cultural version of the [racist] stereotype' (Sanders 2002: 108). But it has had differing interpretations. One that surfaced often was that it meditates on the idea of a culture's destruction by powerful external forces (Nienaber 1966).

Van Wyk Louw was a passionate advocate for an Afrikaans national literature – in the face of what appeared as overwhelming English domination. In addition, he saw radio as a proactive force in the drive for awareness of a new Afrikaans literature and used the medium constantly not only for talks but also for radio plays (Pheiffer 1982; Hofmeyr 1987).[7] Yet he also resented any control of literature and writers by politicians. His 1956 verse drama *Germanicus* focused on the tussle between the pragmatics of power and the intellectual: Germanicus, the brilliant soldier who is also a thinker, resists the urges of his father, the Emperor Augustus, to compromise, and, as a result, dies young, possibly poisoned.

In its Zulu version as a choral music drama, the piece made its own points about community, power, and leadership for its listeners. Two models of a leader are embedded in the original epic poem and in its Zulu 'translation': one is Raka, a powerful and appealing outsider who is a destroyer. The second, embodied by Koki, is the heroic insider who sees danger, confronts it courageously, and is destroyed in the process. In its transformation and translation into a Zulu radio operetta, the strong narrative line of the original remains. So, too, does the contrast between the heroic, insider idealist Koki and the appealing and destructive Raka, with its echoes of the Beowulf–Grendel tussle and prophetic resonances

[7] P. H. Pfeiffer lists van Wyk Louw's published radio plays as *Dias* (1952), *Dagboek van'n soldaat* (1961), *Die held* (1962), *Lewenslyn* (1962), *Kruger breek die pad oop* (1964), and *Asterion* (1965).

of contrasting styles of African political governance.[8] The move from Afrikaans epic poem to Zulu musical drama became an account *from within* of the victory of a dictator figure, the destruction of a community, and the failure of the visionary hero. The warnings against a powerful dictator and the destruction of a small group or nation at the hands of a mightier power were both readings that were also drawn by contemporary Afrikaans critics from the original poem (Nienaber 1966). What was also interesting in terms of its powers of affect and pleasure was the medium. In a way, then, the migrated text of van Wyk Louw began what was to be a new direction on Radio Bantu. In this new space, Buthelezi could begin to craft his own works, drawing on his own archives of cultural knowledge and differing musicalities, always listening to requests from his black listeners. It may not have been a free space but it was one that had possibilities.

As he worked on *Raka* with the Radio Bantu Choir, Buthelezi was busy with something far closer to his own interests and creative centre. *Raka* was broadcast on 4 January 1963.[9] The music drama *uNobathakathi*, scripted and composed by Buthelezi, and performed by the Radio Bantu Choir, was recorded in early February and broadcast in late March of the same year.[10] Buthelezi later remarked that the idea for the production came from a story sent in by a listener.[11] So close was the relationship between listener and the creative figure within the studio, in this case Buthelezi, that requests for repeats were common. Buthelezi's son, Thami, remembered that *uNobathakathi* was 'very popular with some listeners', as were *uNokhwezi* and a later drama series called *Nsakansaka*. 'They were very well known. The listeners would react and there would be a public demand for it to be rebroadcast and the whole series would be.'[12] Again, here we see the interaction of a diverse listening public strung across classes and in both rural and urban areas engaged with the mediating figure of Buthelezi, the focus being on the forum, the *inkundla*, created by the broadcast music dramas themselves.

The bond with listeners

Thami Buthelezi's recollection of the 'public' response to his father's radio works reminds us of the different publics being formed during the early and mid-1960s (Warner 2002). The urban listeners, too, were likely

[8] My thanks to Loren Kruger for fruitful discussions on this and many other points.
[9] SABC Sound Archives for 1963. [10] Ibid.
[11] Remark to be found in an article on the 1965 production of *uNokhwezi* winning the merit award in the radio section of the Prix d'Italia competition. See 'Entry for Italia Prize', *SAUK-SABC Bulletin*, 18 August 1965, p. 18.
[12] Interview with Thami Buthelezi, 2012.

to have come from two socially different groups: first, the aspirant black middle-class teachers, school-goers, nurses, clerks, and clerics who made up the bulk of choral singers and audiences; and second, migrant workers often living in the single-sex hostels controlled by the municipalities, or employed in the mines. Others, mainly women, were in precarious and often lonely domestic employment, and there were those outside even these groups who lived in the interstices of black urban life, who enjoyed the music of the shebeens and bars tucked away in urban streets and township yards. No doubt other listeners were the audiences for the *isicathamiya* concerts that, from 1961 onwards, would find a presence on Radio Bantu. A novel such as *Indlela Yababi* (Path of the Wicked) by R. R. R. Dhlomo (1962 [1939]), who was also a columnist on women's affairs of the heart in the 1930s for the newspaper *Bantu World* (Thomas 2008), captures well the hectic if perilous pleasures of the urban scene in centres such as Durban and Johannesburg. It allows, too, a sense of who might listen to the programmes on Radio Bantu that were to come. Also important for the new station were the huge numbers of rural listeners who would again be teachers, clerks, shopkeepers, and so on, but also the peasant farmers and farm workers in Zulu-speaking Natal.

Certainly the music drama *uNobathakathi* shows that the station's publics would tell producers and scriptwriters in the Zulu-language section of Radio Bantu what they wanted to hear. The bond between listener and in some cases a particular radio announcer or producer was clear. It was not an instruction from above but rather a message coming from within a shared community. On *uNobathakathi*, the producer was Sagius Thwala and Buthelezi himself took the role of narrator; his daughter Harriet was one of the singers. *UNobathakathi*, was, in more senses than one, a 'family' affair. Lyrical, filled with the light soprano voices suggestive of a young woman's social – and metaphysical – world, the music drama provided an entry into the intimate and orderly domain at the heart of a culture, its marriage practices, carried out with absolute regard for established procedure. But it did so only to show how such order is an illusion in the face of the dangerous world of the occult, which exists alongside it. *UNobathakathi* was set in an indefinite yet distinctly historical time: 'long ago in KwaNjabulo, KwaZulu', under a mountain, near a black forest at the place of Ngila'.[13]

There is a touching homeliness in this music drama, the story of a young girl who is kind and generous and the mainstay of her family. The first song sketches in the rural domesticity of the girls going to collect firewood; the solo soprano voice of uNobathakathi urges on her companions, who then join in:

[13] The sound recording of *uNobathakathi*, held at the SABC Sound Archives in Durban, formed the basis of my writing here.

Woza, as'hambe siyotheza

Come let's go and collect firewood

Nomavila!

You lazy thing![14]

Almost immediately, one of the girls raises the topic of uNobathakathi's marriage proposal:

GIRL: When will you answer the *umkhongi* [bridegroom's messenger]?
UNOBATHAKATHI: I'm scared.
GIRL: Why? You'll have your own home.
UNOBATHAKATHI: He's coming tomorrow.

The narrator announces the *umkhongi's* arrival; the barking of the homestead dogs (which becomes a staple indication of a rural home in dramas that follow) welcomes him. In his rich baritone, contrasting with the light sopranos of the young girls, the *umkhongi* lists, enticingly, in rhythmic song, all the cattle that the bridegroom will give to the homestead in exchange for his requested bride. Each beast has its precise name reflecting its colouring, thus bringing out the extensive vocabulary for cattle in the Zulu language. Thus, the *umkhongi's* list includes:

Isithol' enzimakazi, nehwanqakazi ...

A pitch black heifer, a red and white patched one

Nekunz' emdaka ... *nomhlophekazi*

A cream-white bull ... and a pure white animal[15]

The necessary charade of the father's reluctance to acknowledge that he has a marriageable daughter, his pretended indifference to the rich tapestry of cattle put under his nose by the *umkhongi*, is quickly followed by acceptance. The whole Ngila household sings a lively chorus of welcome and approval. The *umkhongi* is delighted with his warm reception and all seems well. Yet it is not. The shadow cast by the very name uNobathakathi (Miss Witches) creeps in. At the wedding celebration, which quickly follows, snippets of dialogue between the men of the groom's party touch on the fact that the bride's father has definitely been seen riding a baboon at night in a neighbour's yard (a sure sign that he is a wizard). 'Does the groom know?' they wonder, adding: 'But apart from that, the bride is very pretty.'

[14] The text is in isiZulu throughout but I sometimes, as here, use an English translation. Unless stated, the translation from isiZulu to English is mine.
[15] See the list in Nyembezi and Nxumalo (1966: 57, 58, 61) for the cattle names listed here.

Witchcraft and the occult become domesticated and woven into the fabric of the everyday, as the little *imikhovu* (wizard's assistants) of the Ngila household hold their own sad, out-of-tune dance, and sing in their distorted, lisping voices (*imikhovu* always lisp):

> *Ngizhogwenzenzhani? Ubani uzhoshiphakela manje?*
>
> Oh vhot am I doing to do? Who'sh dwoing to dish up for ush now?[16]

'Yes,' adds the narrator, noncommittally, as the occult is rendered momentarily comic and poignant, 'the hearts of the *imikhovu* were filled with pain.' The wedding in all its dramatically rendered authenticity speeds on, with its multiple sites of song and action: the young girls sing and sway in their undulating lines, the young men wildly perform the *giya* dance, the women spur everyone on with their appreciative cries, the *induna*'s representative asks the bride and groom sternly: '*Uyamthanda na* [Do you love him/her]?' Radio renders all this 'real', makes it intimate yet distances it. The site of affect and involvement is also one of contemplation.

The focus of the music drama is on a marriage destroyed by witchcraft, and by the 'enchanted world' in which 'subaltern' everyday experience is partially rooted (Chakrabarty 2000: 77). In a section where the narrator's voice, interleaved with minimal dialogue, overlays the choral singing, the impossibility of the bride's new role breaks through. She tries to forget 'all the strange customs [*imikhuba*] of her home', the narrator tells listeners, but one night she is inexplicably woken and feels compelled to go outside. She placates her surprised husband. 'Don't stay long,' he says, 'and look out for witches.' He secretly follows her into the bright moonlight and tracks her path to the river and the graves. Here, she charms the corpse of a man called 'Mbhilakhahlela',[17] and in her sweet voice says, 'My medicines will wake you. Your people think you're dead, but you've been enchanted by me. I want you to go to *us*.' She lifts his body, hides it nearby, and returns home. The horrified husband waits restlessly for dawn when he rushes to report all to his father. As the groom's family gathers, the accusation proceeds and the grisly truth is laid out. UNobathakathi denies all. '*Musani ukunginuka* [Don't smell me out as a witch]!' she cries out piteously. In a dramatic moment, the groom spits at her as he declares, 'I saw you. I followed you. Get the body!' He adds with dramatic finality, '*Phum' emzini kaBaba* [Leave Father's house]!' As song re-enters the narrative, it seesaws between

[16] Normally, '*Ngizokwenzenjani? Ubani uzosiphakela manje?*'

[17] Her calling him by one of his praise names – in full, *Mbhilakhahlela sidl' ihlobo* (literally, 'Maize hurry up and ripen so we can eat you when you're ripe', or 'Hurry and grow into the fine man we can see you're going to be, so we can enjoy you!') – suggests, chillingly, that she knows him as a member of her new household.

the shocked chorus of the accusers and a wider sympathy for her fate. The slow, sad '*Akazi uyoshonaphi*' [She knows not where she can go] is followed by uNobathakathi's sorrowful song addressed to her own father, '*Wangiyala, Baba*' [He has rejected me, Father]. As uNobathakathi sings her unhappy solo, '*Ngibuyile*!' [I've come back], only the little *imikhovu* are overjoyed and welcome her back in a lisping, untuneful song. But her return is an utter disgrace; the marriage is annulled and the treasured cattle are all returned.[18]

Buthelezi here seems to have followed his listeners' anxieties over witchcraft and worked the particular account sent to him by a listener into a modern mini radio opera, one in which he could build on his own preoccupations as composer and innovator and still carry with him his audience of rural as well as urban listeners made up of workers and the fast-modernising African middle class. The focus on a young woman who is not malevolent but is stereotypically, normatively 'good' and simply caught up in her family's fate draws out the fear of witchcraft, of hidden danger in the midst of normality. It also affords a chance for a public contemplation of the precarious situation of many young brides in rural homesteads, where misfortune, construed as the malevolent act of a witch, would often be blamed on the most vulnerable stranger in their midst, the young bride, or an outsider from another lineage. The music drama is thus about danger – the danger *to* women as they move across lineages, and the danger *of* women as perceived sources of contamination and destruction within a lineage. Also, although set carefully 'long ago', it allows, partly through its distancing, a contemplation of the powers of the unforeseen and the unpredictable to destroy happiness, security, and prosperity. The anxiety captured in *uNobathakathi* was an emotional metaphor that was to feature constantly, in various guises, in countless radio dramas in the decades that followed. Voice also functions in a particular way, since it complicates as it foregrounds the subject position of the young and vulnerable bride, *umakoti*, a subject rarely given a voice in traditional discourse on marriage.

Was Buthelezi's excursion in *uNobathakathi* a journey into a romantic ethnicity that played into the hands of the segregationists? Was it the work of a cultural nationalist? Or was it the experimental work of a moderniser taking elements of Zulu culture into a complex and in some ways hostile present? Other imaginative works of the time in isiZulu were tackling 'the modern' through other media and crafting different responses for a black listening and reading public. One such work was C. L. S. Nyembezi's (1961) novel *Inkinsela YaseMgungundlovu* (The Flashy

18 My discussion is based on the recording of the broadcast kindly made available through the Radio Production Service of the SABC with the assistance of Dennis Mseleku, the sound archivist at the Durban SABC, K. E. Masinga Road.

Man from Pietermaritzburg). This became a serial drama on Radio Bantu's Zulu stream in 1963 and ran for 37 episodes.[19] K. E. Masinga voiced most of it, shifting tone, timbre, and accent to capture the sharply contrasting figures of guile and decent hardworking honesty in this classic novel. Hubert Sishi, based in the Durban SABC studios in West Street, was the adapter (Sibiya 2001b). Some sense of the listeners 'out there' at the time comes from the reminiscence of the novelist, writer for radio, and scholar Nakanjani Sibiya. He could recall, in 2012, that as a boy of 11 in rural Eshowe he would listen to every episode of *Inkinsela* and nothing would be allowed to get in the way of that act.[20]

There was thus, in the years of the early 1960s, a variety of feasts for the ear being built up for listeners. Powerful bonds of affect and radio aesthetics were being forged, as were radio communities across spatial divides. The ambitious mini opera crafted in response to a listener's story and the extended serial drama with its ironic take on the flashy purveyor of 'white man's knowledge', who doubles as a modern version of the trickster figure central to so much African oral narrative, can each be seen as voices in a matrix of offerings from within, for a loyal and increasingly 'knowledgeable' audience for an African-language radio station with a dubious genealogy. Voice was being given, and voices distinctive and new were being made.

The hero within: *uNokhwezi* (Morning Star)

UNokhwezi, we know from the testimony of Alexius Buthelezi's son Thami, was much loved by listeners.[21] Partially focusing on the courage of a young woman, linked mythically to the Morning Star (hence her name), and on the strong bonds between mother and daughter, the music drama centres thematically on a male hero figure from within who destroys the aggressor through courage and cunning. It thus stresses agency and success from within, unlike *Raka*, where Koki, the heroic insider, is killed by a powerful outsider. Buthelezi's musical production, broadcast in April 1964, following on the heels of *uNobathakathi*, must have seemed in some ways the perfect 'Bantu' work – timeless, utterly harmless, a neo-traditionalist gem. Yet, as writers on modern nationalism have frequently pointed out, the imagining of the past and the reuse of cultural forms are essential aspects of imagining a nation (Gellner 1983; Chatterjee 1993: 75).

Buthelezi was refiguring a number of motifs in the Zulu *inganekwane* and making something new. Oral narrative – a genre that spread

[19] SABC Sound Archives for 1963. [20] Interview with Nakanjini Sibiya, 2011.
[21] Interview with Thami Buthelezi, 2012.

throughout the Southern African region, often filled with perverse and magical creatures – had sometimes been infantilised by well-meaning missionaries turned educationists who toned down often scatological narratives to produce palatable classroom texts (Scheub 1975; Hofmeyr 1993). Yet the primary readers produced at this time by C. L. S. Nyembezi, for instance, in many cases kept a sense of this wild diversion from 'the real', and in many homes stories were still being told to young listeners at the end of a day's work.[22] The power of the *inganekwane*, as narrative drawing on the fantastical and the melodramatic, with its Manichean world of good and evil (Brooks 1976), was constantly reshaped and reworked by the scriptwriters of the *imidlalo womoya* (serial plays of the air) as they searched for dramatic means to capture the modern. Buthelezi was doing the same, but with music and song rather than dialogue spliced into short units with bridges of sound, which was the typical technique of the serial drama.

The tropes of difficult journeys with the prospect of violence, the unknown, the illusion of surface reality, the need for courage, and possible magical intervention were well-known features of oral narrative – and still are. Buthelezi drew on all these in his new work, which was broadcast in April 1964. The role of the narrator, at that point still important in the aesthetic of the serial and music drama, was played by Gilbert Phewa; this time Buthelezi played the role of the king, uNokhwezi's father. Winnie Mahlangu sang the lead role of uNokhwezi, and other performers who became staple names of the Johannesburg Radio Bantu station included Mary Nontolwane and Stanford Mlaba. In its original version, the one known and loved by listeners, it held the lyric sweetness of *uNobathakathi*, but the voices of the young women were full of jealousy. The enchanting theme song of the young girls, which weaves its way through the whole drama, had words of poison:

> *Nokhwezi, weNokhwezi, umise, umise, umiswe phezu kwethu.*
>
> Nokhwezi, oh Nokhwezi, she's got a place, been given a place way above all of us.

Men have eyes only for Nokhwezi, and all the other girls' suitors are drawn to her. Her beauty, reflecting that of the Morning Star, shining at her birth, makes the other girls all fade into anonymity. They hatch a deadly plot to kill her. Shortly before her wedding to the glamorous uMdlalakwendoda, the girls of her age group invite her to go with them to cut grass in a distant place on the other side of a dark and dangerous forest. Beyond that, and near the best grass, is a deep, fast-flowing river. The spoken voices of the young women convey the thrust of the plot:

[22] Sikhumbuzo Mngadi, personal communication, 25 August 2016.

VOICE 1: *Izilwane zomful' omkhulu zilambile* [The creatures of the big river are hungry]
VOICE 2: *Izilwane zomful' omkhulu aziloboli* [The creatures of the big river don't pay bride price]

Then the opposing strand is introduced; this relates to women's love for each other, specifically mother and daughter. In a dream, an *idlozi lomusa* (kindly ancestor) speaks to uNokhwezi's mother. Warning her of the danger, he instructs her to give her daughter a necklace from the ancestors to take with her on her dangerous travels. Thus the domain of the ancestors links mother to daughter in a protective circle. The necklace, often given as a symbol of love in courtship, draws together mother, daughter, and 'loving ancestor', and so a strand of female power guided by love, not jealousy, is introduced as a counterweight to jealousy and destructive emotions and forces. The journey begins.

As Nokhwezi's innocence turns to a realisation of her companions' evil intentions, the stakes are raised. The precious necklace has come loose from its thread and Nokhwezi has to carry it as beads. The dark forest takes the girls to the home of an ogre, a cannibal *izimuzimu*, and they narrowly escape death. They reach the edge of the forest, beyond which is the deep river, but Nokhwezi discovers that the protective beads from her mother have been left behind. Courageously, she returns alone on her quest to retrieve the necklace. Captured by the bloodthirsty *izimuzimu* who has already killed his *isalukazi* (old lady) in a fit of rage, Nokhwezi is put in a bag and becomes a captive voice as the ogre goes from place to place carrying his singing bundle (*umdumba-dumbane wakhe*); he taps the bag to make her sing, thus earning food along the way. The plotting girls return sorrowfully to the royal place, where they meet a terrible end as their true motive is revealed. After divination reveals that Nokhwezi is alive, and in the power of an ogre, her young suitor uMdlalakwendoda (Best of All Dancers) offers to go in search of her. His difficult journey needs courage, and, the drama stresses, quick wit and intelligence. Through guile he brings the ogre and his *umdumba-dumbane* back to the king's home, where he is tricked and imprisoned. Nokhwezi's mother frees her daughter, filling the ogre's bag with terrible stinging things and the wedding proceeds. The ogre has a suitably grim ending: with his head stuck in the mud as he is pursued by bees, scorpions, and snakes, swarms of bees eventually make a nest in the bones of his buttocks, and that is all that is left of him.

Buthelezi's sung and narrated, highly episodic yet focused story of courage and beauty triumphant in the face of danger and evil forces was without doubt an immensely satisfying musical *inganekwane*. It contained echoes of many tales but was absolutely novel as well. Possibly in terms of its assertion of power through its multivalent performed narrative, its affirmation of agency from within, there was something deeply satisfying about it for listeners battling with the humiliations and restrictions of the increasingly 'Bantu-ised' world in which they were forced to live. The term '[*a*]*bantu*' was being increasingly plucked from its rooted 'people' meaning, dehumanised, and submitted to a 'semantic Nazification' similar to what had occurred with the term 'Jew' in Hitler's Germany (Apter 2005: 74–9). Clearly, in addition, *uNokhwezi* broke new ground in radio form and aesthetics. In terms of its use of sound, besides weaving in the multiple sonic elements of an aesthetic of a performed sociality, it suggested the presence of the natural world in harmony with the forces of goodness. Thus, for instance, it made light, suggestive uses of bird sounds and – as in many folk tales from Africa and elsewhere in the world – at certain moments the hero or heroine is guided by a small animal sympathetic to his or her plight. In *uNokhwezi*, it is the tiny *ingede*, the honey bird, which shows uNokhwezi's lover the path to the ogre's home.

A second version, a prize, a journey

For some reason – again, it needs to be emphasised that the SABC archives are thin on such details – it was decided to make a second version of *uNokhwezi*, for official entry in the Prix d'Italia international radio drama competition. The original script, possibly by Musa Msane in the first instance rather than Alexius Buthelezi, was reworked. Alexius Buthelezi remained in control of the musical score and song; the narrative remained the same, but the whole was entirely changed in some ways, as it was being prepared for a prize and a different audience. Sound as a distinct component within the whole became dominant. All the highs and lows of the first version – the light shifts in focus, the textures of song heard by the home audience – paled into nothing in the second version, alongside the presence of extraordinary sound effects and the overwhelming single voice of the narrator. K. E. Masinga – master of the dramatic and the melodramatic – became the narrator in this version. Winnie Mahlangu kept the role of uNokhwezi, but the choral songs largely faded away. Instead, there was a point and counterpoint of bird sounds, bird calls, animal sounds suggesting things that were largely evil and destructive, and the sounds of the clashing of elements – thunder, wind, and mythic forces set in opposition. All this was not easily done at the time, as sonic effects within a drama were achieved with far more difficulty in a predigital era.

At the time, when the Johannesburg SABC worked from its downtown Commissioner Street offices, with Troy Street around the corner, sound effects were likely to have been made in the basement and the sound mechanics would follow the script and come in at the appropriate moment. Special devices produced specific sounds; wind – and there was a lot of wind in the second version of *uNokhwezi* – was made by the vigorous turning of a sail on a drum. A discussion with two contemporary sound mechanics at the SABC, Schalk van Wyk and Frikkie Wallis, made it clear to me what great effort must have gone in to producing *uNokhwezi* (2).[23] The drama must have been completely reworked, they concluded after listening to each version. In those days, once a drama – or, in this case, a music drama – was recorded it, could not be altered. The jackals, and lions, the multiple birds, including the oriel and the barbet, and small creatures such as crickets and frogs all had a distinct place in *uNokhwezi* (2). Other sound effects made their mark: the narrator's voice – in Masinga's case, distinctive and memorable at the best of times – quavered with emotion and energy; light and dark became intensified through the vocal enunciation as well as in the contrastive sounds. The dream voice of the 'loving ancestor' was given an acoustic effect and had an echoing quiver to it, removing it even more from the material world of the living. In all this, however, the basic narrative and the skeleton of the music remained unchanged. It was, in a sense, simply distorted to achieve a different overall effect. And in some important ways, those effects were remarkable – for instance, the innocence and beauty of the girls' singing set against the fearsome voices and laughter of the *amazimuzimu* (cannibal ogres). Yet, as the audience would realise, this was a false contrast, as the girls were far from innocent.

There were one or two other tiny but interesting details: in the first version, a space of cleared ground at the edge of the forest, deserted because of the ravages of '*izinswelaboya*' (evil people) (Doke and Vilakazi 1972: 773), briefly appeared in the narrative, as the girls first approached it; in the second, this reference was removed. The term '*izinswelaboya*' was one of the many words that the apartheid regime hijacked for its propaganda purposes (Tomaselli, Tomaselli, and Muller 1989), and, in this case, it became associated specifically with those linked to the forces of liberation: 'terrorists'. Even during this early period, only four years after the banning of the ANC and the Pan-Africanist Congress (PAC) in 1960, and with the formation of Umkhonto weSizwe (MK) a relatively recent event, such vocabulary was in operation and would have featured particularly in news and information items. Its presence in *uNokhwezi* is a reminder of intrusive hands even in the dramas.

The second *uNokhwezi*, blending folklore and multiple strands of choral music, with its bold venture into the world of sonic textures within the genre of the radio and music drama, was technically a huge success.

[23] Interviews with Frikkie Wallis and Schalk van Wyk, 2006.

UNokhwezi was submitted in 1965 by the Radio Bantu station of the SABC, marked clearly as being by Alexius Buthelezi, and received a merit award in the coveted international Prix d'Italia, Europe's most prestigious broadcasting competition. A photograph in the SABC *Radio News*, with a short write-up (April 1965), showed Alexius Buthelezi at the Commissioner Street studios in Johannesburg, in conversation with the young politician Chief Gatsha Buthelezi and the Zulu king Cyprian Bhekuzulu kaSolomon. There was, one could say, a certain neat 'Zuluness' in the picture. It was, though, also a 'Zuluness' crammed with ambiguities. The picture fulfilled a stereotype. But the music drama – namely the first *uNokhwezi* – allowed a recuperative and regenerative energy to be transmitted to its multiple audiences that was far from stereotypical. This, perhaps, was the real prize for its creator, Alexius Buthelezi. As *uNokhwezi* (2) slipped away from the SABC and was played on a number of international stations, it became part of a new cluster of 'South African culture'.[24] When it was recorded in 1966 by a listener on SudWes Funk (South West Radio) in Germany, the drama was followed by the voice of Miriam Makeba singing a Xhosa war song and other songs from her first album – none of which were ever played on Radio Bantu; there was also a gumboot dance, a version of '*Nkosi sikelel' iAfrika*' (God Bless Africa), the (national) anthem of the banned ANC and the national anthem of the newly independent states of Zambia and Tanzania, and jazz.[25] *UNokhwezi* (2), heavily folklorised and sonically full, yet still with the shadow of the original, became a sign among a number of significations of an emergent South Africa. It showed, too, the wider link with radio aesthetics, the SABC world of which Radio Bantu was a part, and a broader connectivity with the world of radio. The first *uNokhwezi* stayed 'at home' (*ekhaya*) and made its mark there. In its quest for innovation through song, drama, myth, and oral narrative, and its use of the modality of radio, it became part of a new horizon of expectation set out for listeners *ekhaya*. The mythic, through radio, could be used to recoup the present. Far from being a marker simply of a link to a deep past, myth, as Marina Warner (in Coupe 1997: 189) has pointed out, can be understood as rereading, rewriting, and recreating. From this viewpoint, myths 'offer a lens which can be used to see human identity in its social and cultural context; myths convey values and expectations which are always evolving . . . are never set so hard that they cannot be changed again' (Warner 1994: 14).

[24] 'Second musical by Bantu announcer', *SAUK-SABC Bulletin*, 9 September 1964, p. 9.

[25] Professor Michael Wetzel recorded *uNokhwezi* in 1966 and passed a copy to me in December 2004 at a conference on 'Myth and Media in Africa' at the University of Leiden. I then traced copies in the SABC Sound Archives in Auckland Park and Durban. Peter Raseroka of the Auckland Park Sound Archives and Dennis Mseleku from the Durban Sound Archives gave me invaluable assistance in all this.

Renaissance, recuperation?

Buthelezi's music dramas, which attempted to fuse energies from different cultural performative sites into a single new genre, need to be seen against a long history of earlier strivings by black intellectuals to achieve a new kind of modernity. Earlier writers such as Petros Lamula (1924)[26] and H. I. E. Dhlomo (1985) had engaged in a search for an African present that was expansive and inclusive. It could embrace aspects of the past in a way that avoided ethnographic essentialism, which was celebratory of 'being Zulu' and of being a modern African, the 'New African' espoused by Dhlomo. Lamula had published his cultural nationalist yet pan-Africanist text *uZulu kaMalandela* in 1924, with its underlying purpose of creating 'a myth of national regeneration, ancestry, migration, liberation, a golden age, and of decline and rebirth' (La Hausse de Lalouvière 2000: 105). H. I. E. Dhlomo's essays on Zulu heritage in the Durban-based paper *Ilanga lase Natal*, and his plays on the Zulu kings, showed a similar preoccupation with cultural heritage and reshaping. In 1954, his play on King Dingane (written in 1936) was staged in Durban by the students of the segregated University of Natal Medical School. It allowed a public airing of pride in a national Zulu heritage in a climate of increasing racism and denigration of African achievement by the new Afrikaner nationalist government, which had then been in power for six years (Kruger 1999; Peterson 2000; Sithole 2008). Buthelezi worked through the medium of radio with his own gifts of musicality to bring in new communities of the ear. The figures noted above were part of his artistic and visionary lineage.

Yet radio, repressive as the wider ethos may have been, gave him the freedom to work with other forms of culture more clearly linked to the popular. This space, too, was part of his attempt at regeneration, recuperation, and, perhaps, renaissance. One of the freedoms that working with the medium of radio allowed Alexius Buthelezi and his radio contemporaries was the opportunity to experiment with different combinations or mixes of genre. The writer, working with the producer, could push in new directions, exploit the aural quality of the medium and create new soundscapes that uniquely marked drama production on the station. Buthelezi seems to have done just that, although this was always as part of a team of actors, producers, and writers, some of whom – Winnie Mahlangu and Mandla Sibiya, for instance – seemed able to operate, like him, with equal skill in the different sectors of production linked to radio drama.

A selection of records from the SABC radio drama archives at Auckland Park, Johannesburg, dating from 1963 to 1969, shows Alexius

[26] See La Hausse de Lalouvière (2000).

Buthelezi's versatility. He wrote, produced, acted in, and played the narrator in dramas of different kinds. In terms of an English-derived taxonomy of genres, some would appear more like mini operas. Yet, as an artist and cultural broker, constantly moving between languages and forms of expression, Buthelezi was well aware that such categorising was foreign to cultures where the dramatic and the sung constantly interacted. His musical productions show that this was a period of experimentation in the Zulu programmes. The versatile radio team at Commissioner Street was engaged in a multiplicity of performative sites for the medium. This was part of the search for a modern voice by the writers and broadcasters, both by means of the recuperation of older forms and through the discovery and working out of a popular narrative form for Zulu radio. Buthelezi worked with both the experimental *inganekwane* (oral narrative/musical) and with narratives relating to the preoccupations of daily life; these also often incorporated the life of dreams, voices speaking from beyond the grave, and the extraordinary soundscapes that were soon to become a feature of the distinctive aesthetic of Zulu radio dramas. These, in turn, mirrored an attempt to engage with a sense of the self, of past and present psychically intertwined in a way that was distinctive of a particular kind of modernity being forged.

The SABC Sound Archives – there are no written archives that remain – mark Buthelezi's moves between these two sites of drama production. They also show his involvement in other new radio genres such as the historical drama, which represented another kind of experimentation and regeneration. The multivalent historical drama allowed a mediation through voice of moments of black history set outside the surveillance of the master. Possibly the scope of the imaginative terrain opened up by radio musicals appealed to him most. Yet he seems not to have seen the two forms as mutually exclusive and moved easily between them.

One of the marks of Buthelezi's music dramas was their inclusivity when it came to women. The recuperation he envisaged – and perhaps what his colleagues shared with him – was not one of male heroes. One of the most famous of the early radio dramas was aired two months before the April 1964 broadcast of *uNokhwezi*. This was set in the serial mode of popular radio plays that was largely to characterise productions for the next 40 and more years. Here, too, Buthelezi was involved, this time as producer. The veteran Johannesburg-based broadcaster Mandla Sibiya, remembered by Winnie Mahlangu as 'a very funny man',[27] wrote the script. The serial play was again named after a woman, but it was not a name that had the sweeping cosmic connotation of Morning Star (*uNokhwezi*). It was, simply, *uDeliwe* (Contented One) (see Modisane

[27] Interview with Winnie Mahlangu, 2006.

Figure 2.2 Winnie Mahlangu, first woman broadcaster of the SABC's African-language programmes. Behind her (standing) are Samson Mboweni (Pretoria), Baldwin Langa (Durban), Peter Direko (Bloemfontein), and Sikani Phirinyane (Johannesburg), 1956.
Source: *Drum* photographer, Baileys Archive/AMO.

2013). It dealt with the dangers of city life for beautiful young – and virtuous – women, a theme that linked it to a body of fictional work in English and isiZulu that covered the same narrative terrain.[28] Here, too, Winnie Mahlangu had a leading role.

January 1965 found Alexius Buthelezi producing an 'operetta' by Johann Strauss, translated into isiZulu as *Ibuzwa Kwabaphambili* (The Elders Know Best); again, Winnie Mahlangu featured in the cast, as did the other versatile actress and scriptwriter, Mary Nontolwane. In June 1965, *Buzani kuBaba* (Ask Father) was broadcast; this was a serial drama of domestic tragedy, which set the demands of tradition and the modern in sharp contradistinction, that ran for 40 episodes. Adapted by Winnie Mahlangu, from the original isiXhosa version, it was produced by Mandla Sibiya. Alexius Buthelezi and Mary Nontolwane were among the cast of seven. This impression of a tight-knit, multitalented team is

[28] For instance, Dhlomo's *Indlela yaBabi* (1962 [1939]) and Alan Paton's *Cry the Beloved Country* (1948).

Figure 2.3 Winnie Mahlangu was a force to be reckoned with on Radio Bantu and Radio Zulu. She was also a key figure in the dramas as actor, author, and translator.
Source: Independent Media.

quite dizzying, as the chief individuals concerned simply shifted roles, but each contributed to a new whole. David Hendy (2000) has noted that such seamless overlapping was often a feature of radio production in the United Kingdom and of the medium in general. Yet the ability to shift sites of engagement and juggle skills may have been a special characteristic of the team on Zulu radio, as it attempted to make a free space in circumstances that demanded constraint and sometimes censorship. In the case of the multitalented team at the SABC in the cramped studios and offices in downtown Johannesburg, it seems to have been the fever of desire for artistic production, and often a spirit of resistance, that drove them. They had a vision of what the station could be, despite all its limitations and the restrictions of the ideologues. As Masonto Buthelezi, Alexius's brother, mused, recalling the early days of Radio Bantu and Radio Zulu:

We wanted to keep up with the English and Afrikaans services. We did historical dramas – on 'the heroes of Zulu history' which Hubert Sishi dramatised.[29]

Again, the records show Alexius Buthelezi's involvement with this historical recasting and recuperation, which was also an attempt to keep a

[29] Interview with Patrick Masonto Buthelezi, 1999.

kind of cultural and intellectual parity between the Zulu-language and the two European-language stations. A single airing in 1968 of a two-handed historical drama, written by Thetha Masombuka and called *uMzilikazi*, lists Alexius Buthelezi and Masombuka as actors, with Muzi Khubeka as narrator. The sound records note that it is a 'historical drama in commemoration of Mzilikazi kaMashobane who died on 5th of September, 1868'.[30] Such an act of historical recall was close enough to Zulu history to be seen as 'national'. At the same time, dramatising Mzilikazi's life, both epic and tragic, as founder of a new people, fighter against the Boers and negotiator with the missionary Robert Moffat, could be interpreted as calling up a figure in black history whose sweep of influence inspired memories of past greatness and a vision of an inclusive 'national' or even regional future, quite unlike that which the SABC controllers had in mind.

The desire to have a literary presence commensurate with the works of high culture on the English and Afrikaans stations may have been one reason why C. L. S. Nyembezi's early classic novel of the failed trickster was turned into drama. Certainly, overall, the presentation of radio dramas in the dramatic political period of the first half of the 1960s threw open to listeners the possibilities of what modernity might be. The Johannesburg team also wrote scripts of novels by writers such as Muntu Xulu; one was the 1966 production of Xulu's novel *Ukubuya KukaNsakansaka eDayimane* (Trouble Returns to Diamond Town, i.e. Kimberley), which ran over 15 episodes. This was produced by Alexius Buthelezi and also featured the stalwarts Winnie Mahlangu and Mary Nontolwane, among others.[31] Alexius Buthelezi thus had a key role in what could be seen as an attempt at a renaissance of a modern, eclectic Zulu cultural presence, reaching through the airwaves to a diverse and eager audience of rural and urban listeners. New, self-aware publics sprang into being in the modernising space of this diverse aural medium. The vision of a classic tradition of high and popular culture, drawing on multiple sources, was clearly present: Buthelezi's adaptation into isiZulu of John Bunyan's *Pilgrim's Progress* (*Umhambo lomhambi*) would have been part of an attempt to 'indigenise' radio, making it the medium for a deep and much-loved text such as this, which, as Isabel Hofmeyr has pointed out, had provided a means of debating modern political roles and, like Nyembezi's novel-turned-radio drama, of imagining new modernities among earlier generations of the young elite in the Eastern Cape (Hofmeyr 2003).

After the experiment with *uNokhwezi* in 1964 (and its reshaping in 1965), Buthelezi did not pursue the music drama as a radio medium in

[30] SABC Sound Archives for 1968.
[31] Interview with Patrick Masonto Buthelezi, 1999.

any major way. Apart from the operetta on the life of Johann Strauss he was mainly preoccupied with popular dramas, and, increasingly, with the programme *Cothoza Mfana*, which became a weekly slot for listeners and performers of the *isicathamiya* genre (Gunner 2014). Only in 1976, as the political landscape dramatically altered, did he venture again into the choral drama. In September 1975 his music drama *uChakijane neSalukazi* (Chakijane and the Old Woman) was recorded, and it went on air the same year.[32] Chakijane, the trickster hero figure (Callaway 1868), has links to other tricksters in Hausa oral stories, and to the epic hero Sunjata, founder of the old kingdom of Mali. He leaps from his mother's womb and is immediately bold and active. '*Nginike ububende* [Give me tasty blood-mince]!' he demands of his surprised mother as she offers him her breast. Fearless, unpredictable, compassionate – he helps an old woman and by doing so shows his humanity – he was a compelling mythic and folk figure for the coming year of the Soweto revolt, 'of fire and ash'.

Perhaps the choice of the trickster figure who crosses between the zones of the dead and the living, with the power to exist outside the normal social structure yet able to return to it, was a fitting choice for Buthelezi's last music drama. It was possibly a final attempt by Buthelezi to remind radio listeners of figures in cultural memory who had always eluded power and domination. Alexius Buthelezi's personal history was not untouched by the fast passage of politics in the later years of the 1970s. In 1978 he chose to leave the SABC and work in the mining town of Welkom, in the Orange Free State, as it then was, as publicity officer for the newly formed political party, Inkatha, with Chief Gatsha Buthelezi at its head. Inkatha and the ANC in exile had maintained close links until 1979, when they split those ties and Inkatha changed from a partial ally to a violent opponent of the exiled ANC. Alexius Buthelezi, as an Inkatha man, most likely became a target for angry ANC supporters. When his house in Soweto burned down in 1981, his papers were scattered all over Diepkloof and beyond and lost.[33] The SABC Sound Archives remain the main repository of his work, with some copies of scores saved in the archives of the Southern African Music Rights Organisation (SAMRO), the non-racial copyright organisation founded by Gideon Roos after Hendrik Verwoerd summarily sacked him from the SABC in 1961.[34]

[32] SABC Sound Archives for 1975. [33] Interview with Thami Buthelezi, 2012.
[34] SAMRO documentary: 'SAMRO 50 years'.

Part II

Distance and intimacy

3 Exile

Bloke Modisane and the BBC, 1959–87

Deadwood Dick: No not that one ... That other one ... the one of the revolution ... You know, the one of Marlon Brando. The one of the Romans.[1]

Modisane and black modernity

On 27 January 1969, Martin Esslin, head of BBC radio drama, wrote to his team of script editors at London's Broadcasting House in Portland Place. The topic of his memo was the script of a radio play by the writer and actor Bloke Modisane. Like its author, Esslin was an outsider. An East European émigré, steeped in the avant-garde dramas of the continent, now in London and a key figure at the heart of a cutting-edge media genre, he could smell a good radio play a mile away.

One of the queries from the team in the script unit concerned the way in which Modisane's central character, Caiaphas Sedumo, spoke. He didn't sound African enough. His accent could have stepped out of the pages of the London *Times*. Esslin was swift in his response. He gave the play his strong support and assured his team that '[e]ducated Africans in South Africa do sound like the *Times* editorials'. He continued: 'I know of few plays which present the predicament of intelligent Africans in Johannesburg with greater truth and compassion.' He also commended the play's brilliant handling of 'the cinematic' aspect of radio.[2]

After that, the future of *The Quarter Million Boys* was assured, and in the systematic way of the BBC radio drama unit, plans for its production began to unfold. Its journey as a script had begun in early April 1968, when the producer Charles Lefeaux wrote a polite note to its author: 'Thank you for sending me *The Quarter Million Boys*.' On the same day, he noted to a colleague: 'offer to third [Third Programme] – is too long of

[1] From *The Quarter Million Boys* by Bloke Modisane (1969), p. 20. Script in BBC Written Archives (BBCWA), Caversham, ref. DM795H.

[2] Martin Esslin correspondence, BBCWA, 27 January 1969. Modisane File 2, 1968–72.

Figure 3.1 Bloke Modisane with telephone, *Drum* offices, Troy Street, Johannesburg.
Source: *Drum* photographer, BAHA/AMO.

course but I think I could knock it into shape'.[3] It was, in fact, Lefeaux who produced it and who would often speak up for Modisane's work. *The Quarter Million Boys* had its first airing (of several) on the 'Third Programme' (later Radio 3) on 26 September 1969.

Esslin's letter neatly encompasses two themes of this chapter: the innovative possibilities of radio as a dramatic medium; and being African in Britain, and in the world. Both these concerns were also central to Bloke Modisane, exile, artist, and black diaspora man. I explore in this chapter Modisane's engagement with the medium of radio, and, more specifically, how he operated as both exile and diasporic figure, mainly in London, then briefly in Rome, and finally in Dortmund, West Germany,

[3] BBCWA, Modisane Scriptwriter File 2, 1968–72.

always maintaining his links with the BBC, particularly with the radio drama script and producers' department at Broadcasting House, Portland Place. I argue that Bloke Modisane's fragmented career as an artist working across a number of media both marks the state of exile and offers new insights into his life as an artist and into the black diaspora of his era. It also offers insights into the mediated voice itself as a site of historical awareness of black history and empire. The names of his fellow artists from South and West Africa and the Caribbean – Peggy Phango, Joseph Marcell, Zakes Mokae, Cosmo Pieterse, Lionel Ngakane, Alton Khumalo, Christopher Asante – slip briefly on to the pages of the BBC's *Radio Times*. If they worked for African Theatre at Bush House on the Strand in Central London, their names are even harder to find. They provide a small window into the world of black actors and writers who were resident in London, or passing through the city, in the 1960s through to the 1980s and who were part of 'black London'.[4]

Modisane is known as a key figure of the South African *Drum* era of the 1950s and the author of an acclaimed autobiography, *Blame Me on History*. This was published in London and New York in 1963 but banned in South Africa, where it was not available until 1986, the year of his death. The trajectory of his life after he left from Park Station in downtown Johannesburg in April 1959, with London as his secret destination, until his death in Dortmund, West Germany, in March 1986 is not always clear. It was a life of fragments – in essence, a perfect 'modernist' life. Unlike eminent figures of the Caribbean diaspora such as C. L. R. James, who ended his life in London as an adored and feted luminary of the Left and of the black diaspora, or V. S. Naipaul, who became a respected figure of the literary establishment, few South African intellectuals from a similar period have left much discernible mark on the cultural or literary memory of Britain's capital. Yet a number of them – and certainly Bloke Modisane – were artists whose London presence complicated and even shifted the cultural sensibilities of the time, particularly with regard to race, blackness, and modernity. They lived as outsiders marked by their colour, many charting a new sensibility of blackness at a particular moment in late empire. Modisane was one of these; others included Sam Selvon, the Trinidadian writer, and the Jamaican cultural critic Stuart Hall.

Modisane played his role largely through the BBC and its radio drama, but he was also in demand as an actor. From the middle of the 1950s,

[4] For instance, the cast list for his 1987 radio play *The Somebody* included Joseph Marcell as the elderly politician released from jail, Kwabena Manso as the restless young would-be people's hero Kalamazoo, the star of *King Kong* Peggy Phango as Kei Amatola's wife, Providence, and Christopher Asante as the subtle new breed of post-apartheid politician, Duke Macharudede.

radio, once the queen of public media, was a lowly second to television (Ash 1985: 3–4; Crisell 1986). Radio drama was forced to assess 'what it could still do better than any other dramatic medium' in its fight to survive. It was on the cusp of this period, when radio drama had found a new and loyal audience, and 'had achieved a continuing level of excellence of a whole corpus of accomplished works . . . broadcast to an expectant and loyal audience of a million or more' (Ash 1985: 3–4), that William Bloke Modisane's first radio play was broadcast. The number of listeners for a single radio play was estimated to be 'some three quarters of a million for the Monday Play . . . with its repeat on the following Sunday afternoon, and about a million and quarter for the Saturday Night Theatre, with its repeat on Monday afternoon' (ibid.: 3–4). In two airings, a play on radio could thus have the same numerical audience as a stage drama might achieve if it played to packed houses over a period of two years (ibid.: 5).

Modisane was not simply one of an elite group whose plays made it through the siftings and differences of the script unit. He was a black writer at a particular moment in British late imperial history, and a South African exile. In some moments he felt the BBC boxing him in. He could say in 1975, when he was admired as 'an excellent writer' by the exacting team in the script editors' unit: 'Do you think I could be considered for commissions other than those of themes dealing with Blacks and the black experience? The field is too restricted. I can do other things.'[5] There was also the question of how to handle questions of nationalism. Paul Gilroy has claimed:

> the problem of weighing the claims of national identity against other contrasting varieties of subjectivity and identification has a special place in the intellectual history of blacks in the west. (Gilroy 1993: 30)

For Modisane, this was combined with his sense of an African-American/African black modernity that was part of a global awareness of colour and black history within empire. Perhaps what Modisane became an important part of, as he wrote his bundle of influential plays for the BBC, was a complicated ideology of voice that at times fitted into this global awareness and at other times traced a black modernist artistic sensibility.

Certainly, for Modisane, the claim of national identity was a pressing question, as the armed conflict within and outside South Africa increased in magnitude. This became even more the case in the 1980s when resolution seemed in sight. Yet he also had affinities with the category of black writers who:

> held out against the closed shutters of 'black particularity'. . . and opted instead to embrace the fragmentation of the self (the doubling and splitting) which modernity seems to promote. (Gilroy 1993: 188)

[5] BBCWA, Modisane Scriptwriter File 2, 1973–79.

Particularity and dialogues of creativity

The question of black particularity presented itself forcefully in the early 1960s when Modisane's work had a place in some of the foremost journals of the civil rights movement in North America; at least one, *The Atlantic*, had a long history of involvement in black rights stretching back to pre-abolition days. Important as London and the BBC may have been for Modisane as an artist, exile, and diaspora man, there was also the ferment of the American civil rights movement, and issues of black identity and art in North America. Even before the publication of *Blame Me on History*, Modisane's work was to be found in journals such as the *Negro Digest*, the *Atlantic* and *The Reporter*. The *Negro Digest*, which published his essay on black short-story writing in South Africa, 'The rage that racism bred', in October 1962 (Modisane 1962) noted: 'We have Langston Hughes to thank for introducing us to this fine writer.' They alerted readers to 'a poem by Mr Modisane' which was to 'appear in an upcoming issue'.[6] This was the poem 'Lonely', which came out in the November issue, and dealt with issues of the outsider, blackness, and the artist, in the style of a jazz-like blues song:

> It gets awfully lonely
> lonely black; like screaming,
> screaming lonely . . .
> I talk to myself when I write,
> shout and scream to myself; then back to myself
> scream and shout.[7]

Langston Hughes was to publish it soon afterwards in his anthology *Poems from Black Africa* (1963). It was to Hughes, who had had links to Modisane since his *Drum* days, that Modisane could pour out his heart about loneliness, play with ideas of black invisibility, complain of money and the insecurity of London exile life, and also chat about the ups and downs of the South African musical *King Kong* on its London run (see also Chapter 4). In the letters he wrote in 1961 and 1962, when Hughes was back in New York and had passed through London, Modisane could let loose about his depression, mock himself, joke about racial slurs, and still talk seriously about his writing. He ended a letter in May 1962:

> I don't know what I want to do, I'm not even sure I wanna do it – if and when I find out. All I know is I'm tired. Nothing is good enough; things don't change, it's all a drag . . . This letter is getting to be a drag, I better cut the commercials – maybe I'll feel better tomorrow, I'll write you a less dizzy letter.[8]

[6] *Negro Digest*, 1962, 11 (11): 97.
[7] Bloke Modisane on African poetry in *Negro Digest*, 1962, 12(1). Available at onlinebooks.library.upenn.edu/webbing/book (accessed 8 August 2014).
[8] Modisane/Hughes correspondence, NELM, 11 May 1962.

Figure 3.2 Langston Hughes (1902–67), a key figure in the Harlem Renaissance, poet, dramatist, and anthologist. Hughes was also a friend of Bloke Modisane and a supporter of his work.
Source: AAI Fotostock.

More seriously, a letter only a few weeks later introduced his writing commitments, and his regard for Hughes:

> You have always made me feel that maybe something good will come out of me yet; when I think I'm going mad, you tell me: nuts.[9]

Some six months later, after they had met again at the African Writers Conference in Kampala[10] and were once more on different continents, Modisane could write:

> I've seldom been hungrier in my life, not even in South Africa ... Anyway, I'm writing my book [*Blame Me on History*] and if I didn't have to dedicate it to my mother it would have been for you. I'm frightfully proud to know you. Just keep believing in me, maybe some day I'll justify it.[11]

Perhaps partly through recommendations by Langston Hughes, but also because of an awareness of his searing testimony of blackness, *Blame Me on History*, which had been published in parts in *The Atlantic* prior to its North American publication by Dutton & Co., Modisane went on a

[9] Modisane/Hughes correspondence, NELM, 31 May 1962.
[10] Modisane/Hughes correspondence, NELM, June 1962.
[11] Modisane/Hughes correspondence, NELM, 21 November 1962.

lecture tour of the US in the early 1960s, not long after K. E. Masinga's American travels.[12] At the invitation of the United Negro College Fund and the American Society of African Culture, he visited and lectured at 'ten Negro colleges in the South, and one in New York'.[13] Back in London, he duly reported on his travels, including commenting on the civil rights movement now gathering momentum, in at least one radio interview, with Lewis Nkosi, and freely gave his thoughts on 'Africa and the American Negro' (see Chapter 4). Clearly, Modisane, the black artist, man of letters, and commentator, particularly on South African race politics, had a North American 'Negro' presence, a bold public persona, at the time. He could also open up his private anguish to a black fellow-writer and discuss with him the intimacies of a life in art.[14] His BBC presence was different but contained its own undeniable potency, that of the searching presence of the black outsider, and artist.

Intimacy, distance, 'a room'

It was as a writer of radio dramas that Modisane was most valued and respected, and his most intimate working relations as an artist were with those in radio drama at Broadcasting House. This close-knit and demanding working environment to some extent created a sense of 'home' for Modisane. It provided 'the room' that could be his creative centre in the medium of sound, drama, and words, even when his life of exile was a maze of restless movement and uncertainty. Far more than simply a series of interactions, there was the making of bonds between those within BBC radio drama, particularly the script unit, and the writer who was without. Modisane's position as a South African and a black exile marked him as an outsider, but increasingly an influential one. Yet the script unit for radio drama was itself a base for gifted outsiders who were often artists themselves. Besides Martin Esslin, there was William Ash, who became a discerning advocate for Modisane. A Texan by birth, from Dallas, Ash served in the Air Force in World War Two then became a communist, moved to Britain, and became an External Service reporter in India for the BBC. Refused entry into the British Communist Party on his return to Britain, he joined the smaller Marxist-Leninist Party, and, after losing his job with External Services, moved to the BBC's Radio Drama script unit. Ash, Charles Lefeaux, and later Bernard Krichefski and Shaun McLaughlin were among those who handled Modisane's scripts and sometimes saw them through to broadcast. The unlikely medium of the departmental memo often showed the long toing and froing of a work before it was deemed 'ready for production'. This

12 *The Atlantic Extra*, November 1963, pp. 96–119.
13 African Writers' Club tapes, C134/454 NSA.
14 The term 'Negro' was used to affirm black dignity in the civil rights struggle and earlier.

happened even when a play was commissioned, and, as Modisane's reputation at the BBC grew, work was indeed commissioned from him.[15]

Modisane's involvement with the BBC predated his departure from Johannesburg and in a sense outlived him, as his last play was broadcast on Radio 4 a year after his death in March 1986. The records of his early years (kept meticulously by the BBC)[16] show him appearing as a public presenter on African issues, but also as an actor, as much for the BBC World Service and its African programmes from Bush House as for Broadcasting House. For Bush House, he was particularly active in the African Theatre programmes, as an actor and sometimes as a dramatist. In a few cases, the African Theatre from Bush House took a play turned down by the script unit at Broadcasting House.[17] He had thus a sort of double voice: there were the plays for the British 'home' audience, from the BBC at Broadcasting House, and the plays from the BBC World Service at Bush House, which were beamed to African listeners on the continent itself. In both voices he was at various times commentator and actor as well as dramatist. Each of these indexed the outsider and the man of knowledge and signalled to wider arcs of global black connectivity and the voice of the black modern.

The links with the BBC began while his address was a Johannesburg one – the downtown *Drum* offices at 15 Troy Street, which were very near the back entrance to the SABC used by Alexius Buthelezi and the team of black radio artists. His first recording for the BBC was a features piece, 'Talking to Africans', which went out on the 'Home Service' on Friday 4 July 1958.[18] The producer was Sasha Moorsom. Beamed to 'home' listeners from stately Broadcasting House in Portland Place, it was part of the drive by forward-looking producers within the BBC to alert its listeners to the new postcolonial cartographies of Africa and Africans that were shaping themselves at the time. The piece must have been a success, as it signalled the beginning of Modisane's long links with broadcasting in Britain, and particularly with radio drama.

The recent or imminent independence of previously colonised countries such as Ghana, Kenya, and Nigeria, as well as the shock of the Sharpeville massacre in March 1960, meant that there was now a place for the black African voice in Britain's capital. Services to the Caribbean had a longer history. BBC radio had opened up its ambitious Africa Service, part of the World Service that operated from Bush House, and the capital was alive with the buzz of Afro-Caribbean writers as well as African writers and politicians. The BBC found a place for many of these voices on its programmes beamed to the African continent. The scene was one of intense activity for Modisane. He worked frequently for the African Theatre programme of the World Service as an actor in radio plays. In addition, he participated in a variety of talks on life in Africa, on travel in Africa, on

15 BBCWA/BBC Sound Archives, Modisane Scriptwriter File 2, 1973–79. 16 Ibid.
17 Ibid. 18 BBCWA Modisane Artists File 1, 1958–62.

South African theatre, and on African literature. Apart from the cast lists of his own plays, which give a clue to the mix of South and West African and Caribbean artists the BBC could draw on, the range of names called upon to discuss and talk on African affairs and African writing suggests a similar energy of outsiders. He was part of the very active milieu of black artists from Africa and the Caribbean who were in London in the late 1950s and into the 1960s: Andrew Salkey is mentioned in one programme; on one occasion he shares an interview with the writer Noni Jabavu; he reviews Ezekiel Mphahlele's *African Image*; and he is regularly part of Edward Blishen's ground-breaking programme for African writers and intellectuals.

He was also acting on the stage. A letter dated 24 May 1961 from Veronica Manoukian, a producer for the African Service at Bush House, refers to his leading role in Jean Genet's *The Blacks* with the Royal Court Theatre, first at Cambridge and then in London. It gives a sense of the hectic life of Modisane at the time, successful as an actor and in demand at Bush House for broadcasting.

Dear Bloke,

I was delighted to read in the Guardian of the success of 'The Blacks' at Cambridge and particularly to see that star of screen, television and radio, Mr Bloke Modisane had such a fine write up. I shall lose no time in coming to the Royal Court myself to cheer you on

...

Could you be kind and ring me when you get back to London and we will fix recording times all over again. [Manoukian was keen to reschedule a talk by Modisane on Kwela music and its recording for the programme 'Soundings'.] I am sorry about all this – this is what happens when I rush into action too quickly – in the future I will go at my normal slow pace.

With best wishes for the rest of the tour,

Yours ...[19]

Modisane was also attempting to have his own fictional writing aired through broadcast; another letter of the early 1960s mentions that a bundle of his short stories was being returned to him. Once he missed an appointment for a schools programme on travel and caused havoc in the tightly run schedules of the recording room. The payment for such work was modest – usually 4 to 6 guineas, with further payment depending on whether the programmes were reused on another network or abroad.

Not the African image?

Bloke, you didn't begin writing in this way did you? Your training was that of a journalist. How has your writing developed into this, into this rather sophisticated form? Can you trace back the development?[20]

[19] Letter from Veronica Manoukian to Bloke Modisane, 1 May 1961; BBCWA, Modisane Artists File 1, 1958–62. A later letter sets the recording for Tuesday 20 June at Bush House.

[20] NSA, African Writers' Club, BBC African Service, Edward Blishen to Bloke Modisane, 27 September 1963.

It is in late September 1963, shortly after the Thames and Hudson publication of *Blame Me on History*, that we first catch a glimpse of Modisane speaking 'in his own voice' as a writer to the presenter Edward Blishen in a programme on African writers.[21] The interview with Blishen contains hints of the 'beggar dependency' of black intellectuals on white readers of the time, something that Mark Sanders has discussed (2002). Yet, it shows how Modisane eludes being pinned into the category of the 'emergent' African writer. Modisane's relish for a modernist form of broken time dominates much of the interview. A feature of the recorded interview is the contrast between the measured, leisurely Englishness of Blishen's speech and the breathless, staccato speed of Modisane's not quite equally 'English' diction. Once again, subtly, voice becomes an index of origin and of a particular place in empire and colonial history:

BLISHEN: Tell us about the writing of the book.

MODISANE: Well, this is where I encountered my initial problem. This is where the problem started. [Laughs.] You see, I had to take a backdrop and I took the breakdown of Sophiatown ... [M]y connection was in the very breaking down of the time. If I broke time, I thought, then I could actually get the time to move back and forth and keep on coming in where I broke off, you see. Of course, the working of this was difficult and I remember after about a month of this I went to see a film called *Last Year in Marienbad* [directed by Alain Resnais].

Later in the same interview, Modisane mentioned Kafka; Blishen's remark on 'the sophistication' of *Blame Me on History* spurred him to expand on Kafka and the importance of the radical Vanguard Bookshop in downtown Johannesburg:

MODISANE: I went to work in a bookshop ...

BLISHEN: The one you mention in *Blame Me on History*?

MODISANE: Yes. And now here I was exposed to the masters of the craft by studying the German, the Russians, the French masters – and seeing how they worked, what they did with time, and when I looked at Kafka – Kafka I think was the first man that really made me realise what can be done. Because this man could create a nightmare – a nightmare, in order to be that, must be in sequences of time that are completely uncoordinated and unrelated. So that when the man in the castle sits there waiting for the orders to come from the castle and nothing comes from the castle; rather, other things are insinuated into his existence; there are sequences of time ...

Modisane concluded the interview by mentioning Athol Fugard's theatre work with Lewis Nkosi and himself, and the crucial role of the alternative and oppositional cultural space of Dorkay House in downtown Johannesburg when Nkosi and he were still based there and were living

[21] Ibid.

in Sophiatown.[22] Through the interview, Modisane was thus refusing the 'emergent writer' slot and creating instead a meditation on modernist cinematic techniques in his autobiography. He was also gesturing at the whole range of his skills, which included theatre and film, and the origin of his work as an artist, a writer for *Drum* magazine, a Sophiatown literary figure sketched into momentary life by writers such as Can Themba – in short, a character in his own right, in the vibrant and oppositional world of black Johannesburg in the 1950s (see Patel 1985). Theatre began for him through his early experimental work with Fugard in his first play, *No Good Friday*, and film and acting through his collaboration with Lionel Rogosin in the secretly made *Come Back Africa*.[23]

Time and sound were key elements in Modisane's first radio drama, *The Quarter Million Boys*. In one sense, there was nothing new about *The Quarter Million Boys*. It was merely an adaptation of his short story 'The situation', which *The Negro Reporter* had published in December 1963,[24] and the Nigerian journal *Black Orpheus* in Ibadan at about the same time. Gangsters who dressed like Frank Sinatra and relished spoken Shakespeare, popular black urban music, and anguish were all marks of the urban culture of Sophiatown – and all of these contribute to the dense soundscape of his first radio drama. But everything was different because it was a new medium, a new genre, and a new time and place. It was, through radio, briefly, a London sound. Modisane's status as an actor, often playing the main role in other radio plays for the BBC,[25] meant that he was in a strong position to produce his own work and move from simply performing what others had written to making his own statement in the medium. *The Quarter Million Boys* – the title being the name of one of the gangs who meet in the Sophiatown shebeen [bar] where the drama is set – was the most successful of his plays. It was re-broadcast many times on both Radio 3 and Radio 4, also aired as part of the Bush House African Theatre programme,[26] and was put out in translation on West German Radio (WDR) only months after its 1969 debut.

The Quarter Million Boys centred on the black intellectual, 'The Situations' as they were called, in the 'dead city' of Sophiatown, negotiating a space for art and survival, caught between demands for easy

[22] Ibid.

[23] This refers specifically to the famous shebeen scene in *Come Back Africa*, in which Miriam Makeba and Lewis Nkosi also feature (see Modisane 2013).

[24] *Negro Digest*, 1962, 11(11): 97.

[25] Charles Lefeaux, a great supporter of Modisane's BBC plays, produced *By the North Gate*, a novel by Gwyn Griffin adapted by the author for radio in 1963. On 22 June 1964, Modisane also acted in *Brother Man* by Roger Mais, adapted for Radio 4 by Sylvia Wynter; he also appeared in *Back into the Light* by James Carey, adapted for radio by Peter Cotes and produced by David H. Godfrey.

[26] This was first produced with a different producer, John Gordon.

violence and for total subservience. The outsider figure, Caiphas Sedumo (played by Modisane himself), is taunted first by aggressive rural white farmers, and second by the black gangsters with whom he shares the space of the shebeen, in the 'dead city'. In each instance, death is close. To both groups he is an unbearable anomaly, either as an overeducated 'Bantu' or as a fellow black who does not 'resist' – although gangsters, as history shows, were unreliable allies in the apartheid struggle. Placed under unbearable pressure by the gangsters, who want at least powerful words, and who want Shakespeare's words on revolution, Caiphas publicly recites the 'Friends, Romans, countrymen' speech from *Julius Caesar*, changing phrases to make the text refer to John Vorster's South Africa, not Rome. He momentarily ignites his drunken shebeen audience to a state of revolutionary fervour and then pulls them out of the illusion:

CAIPHAS: Fellow countrymen, Sons of Africa, hear me ... You don't understand ... This is the cinema ... Nobody really gets killed. Hear me countrymen, stay countrymen ... You don't understand this is the cinema, nobody really gets killed.

DEADWOOD DICK (OFFSTAGE): We'll burn their bodies in Joubert Park [a famous central spot in downtown Johannesburg].

The shebeen queen, Battleship, has to dampen it all down with a sudden twist of tone:

BATTLESHIP: That's enough you! GENTLE ROMANS SIT DOWN AND SHUT UP.[27]

What dominates in the 'blind' medium of radio is sound, and the play used it with all the finesse of the best plays of the period. Lefeaux, who produced it, had won the Prix d'Italia a few years earlier for his production of Tom Stoppard's *Albert's Bridge*, and was, like Modisane, a master of his craft. It was unmistakably about race and power in South Africa but also about rhetoric, and it used what is probably the most famous speech in English poetry, the one recited endlessly in classrooms throughout the imperial world as English spread with empire across the globe: 'The one of the Romans.'[28] It was also, as we have seen, the only snippet of Masinga's translations of Shakespeare extant at the SABC, and the one he recited with revolutionary vigour for his English Service interviewer (see Chapter 1).

The play was full of dialogue that captured the light and dark of the shebeen space and the dramas of difference enhanced through accent. Caiphas, 'The Situation', did not *sound* like the gangsters from the two rival gangs, the Americans and the Quarter Million Boys. Music was integral to the play's sound texture: the shebeen space and the era of the

[27] *The Quarter Million Boys*, p. 23. [28] Ibid., p. 20.

lost Sophiatown were caught by the music. Some, like the penny whistle, was specific to the era of the 1950s, while the jazz – and, in particular, the saxophone – that drifted through the play evoked a free urban space outside any particular milieu, but also a free space of blackness, and the black voice as creative figure within the BBC.

Scenes shifted swiftly, cutting across time, catching suddenly the rough language of the white farmers who had earlier surrounded Caiaphas in the city:

PIET: Go on, speak again. Just one English word. One.
JAKOB: Off with that hat! A white man is talking to you.[29]

At the high point of the play, as Caiaphas's adaptation of Brutus's speech reaches its climax, gunshots, marching, and shouts of 'revolution' dominate and the listener is briefly, like those in the shebeen, within the revolutionary moment.

All this, along with the evocation of a contemporary Western style of dress (Pontiac Mick, the gangster, 'pulls the glove from his hand' and his gangsters are all 'dressed up like Savile Row', smoke Bond Street cigars, and wear Pierre Cardin ties), the evocation of a shared idiom of film (the romance of Charles Boyer, *The Guns of Navarone*), and a seemingly throwaway reference to something as arcane yet current as Ernest Hemingway's treatise on bull-fighting and the nature of writing, *Death in the Afternoon* (1932), gave the play a dense auditory richness that transgressed any simple stereotypes of the African and briefly 'quoted' Sophiatown's lost urban popular culture (Stein and Jacobson 1986). The early puzzled remark that the central character did not sound African enough in fact summed up the startling cosmopolitanism of the play. It was transgressive and fascinating. Modisane had translated the 'global ecumene' of Sophiatown into the medium of British radio; Shakespeare, *Julius Caesar*, and Marlon Brando became the property of characters in a lowlife Johannesburg shebeen; and the angst of the black intellectual was bared, but not solved. Moreover, English itself, as a language of 'the other' as much as of the native English, was a character in the play: two very normatively 'English'-sounding characters, Anthony, an architect, and Rosemary, just arrived from London, appear briefly in the shebeen as interested onlookers and sound endearing, a little naïve, and faintly ridiculous. Here again, the extraordinary range of voice – in terms of accent, timbre, enunciation, and political possibility – itself provides an auditory lesson of a cultural mix – familiar in Sophiatown and parts of South Africa but not, widely, in London.

Finally, after the return to a kind of normality, the listener is left with the voice of a singer cutting across the heightened unease, singing 'The

[29] Ibid., p. 8.

Figure 3.3 Bloke Modisane standing far right, with the BBC radio drama producer Charles Lefeaux and the cast of *The Quarter Million Boys*, London 1969.
Source: BBC Picture Library.

blues is the black of the face', Modisane's poem set to a blues tune.[30] It was a poem that Langston Hughes had anthologised in his *African Treasury* (1961) and which held echoes of Hughes's own blues poetry with its deep undertows of racial pain and defiance and its own radical textuality. Its use in the play was perhaps a form of tribute to Hughes, who had died two years earlier.

Dialogues and distant ties

Modisane's success with *The Quarter Million Boys* marked the beginning of a close collaborative period with the BBC radio script unit in Broadcasting House. Ash and Lefeaux in particular were intensely interested in what he produced, but nothing was accepted automatically and no magic doors were opened. Clearly, though, Modisane's voice had made its mark. *The Quarter Million Boys* itself was in demand and was chosen as one of the plays in a series 'From the Sixties' to be broadcast early in 1970 on Radio 4's Afternoon Theatre slot. In some cases, he worked across the two channels of Bush House on the Strand, with its world and Africa focus, and Broadcasting House in posh Portland Place, with its eye

[30] Ibid., p. 24.

on the mainland and its 'English' audiences. But there were differences. Broadcasting House had its 'home' listeners in mind and sometimes Modisane's work was deemed not appropriate: how, for instance, could begging ever be seen as a profession? When Modisane turned his ironic and wildly extravagant short story 'The dignity of begging' into a radio script, only the Bush House producer, Gwyneth Henderson, took it for the African Theatre series, where it was broadcast twice in December 1971.

Yet shortly after that, a further radio play, a satire on South African political life, *Freedom of the Vices*, was aired after a careful period of gestation between the writer and the script unit editors. Ash asked Lefeaux to 'cast a producer's eye over it' and come up with 'some leading questions which can then be put to Bloke'. The script came back and again Ash wrote, 'I have got Bloke to rewrite this in line with your suggestions. Do you think it is OK now and would you like to produce it?' Even after the copyright form was cleared, the honing continued: could there be less explicit social criticism? 'It might be useful for you to work this out to some extent yourself,' Modisane was told.[31]

At roughly the same time Modisane sent in another play, 'School for Exiles'. On this occasion it was Lefeaux who wrote to a colleague, 'Bloke Modisane has sent me another play which I like very much – it is funny, touching and poetic, with racy dialogue and something to say about the uprooted coloured man's attitude . . . perhaps you could find time to read it and then we could have a word?'[32] In this case there was a lack of consensus and the play was turned down. This pattern of success and some failure after intricate conversations between script editors continued, but Modisane did not leave the difficult conversations with his demanding gatekeepers. His adaptation of the mystery thriller *Many Brave Hearts* by Donald McNutt Douglass, set in the Caribbean, was snapped up – but carefully so.[33] After being commissioned to complete it for the prime Saturday Night Theatre spot on Radio 4, he was still asked for revisions by Lefeaux. When these duly came in, the reply was almost enthusiastic: Lefeaux was going on a week's leave, but was 'looking forward to reading this latest version'.[34]

The play went out in its prime slot in February 1973, but by then Modisane had moved to Rome with his new wife, Ingeborg, having sent Ash a jaunty postcard some months earlier: 'Bloke and Inge Modisane are alive and well and living in Rome.' The postcard, with its new address, was duly filed in the Copyright Library by Ash's secretary.[35] Did the departure from London mean the end of the writer's relations

31 BBCWA, Modisane Scriptwriter File 1, 1968–72, Ash to Lefeaux.
32 BBCWA, Modisane Scriptwriter File 1, 1968–72, Lefeaux.
33 BBCWA, Modisane Scriptwriter File 1, 1968–72. 34 Ibid.
35 BBCWA, Modisane Scriptwriter File 2, 1973–79.

with the BBC script unit? Not at all. After all, it was a dialogue that had to continue if Modisane was to have his voice heard and his art recognised, and make a living. By this point, it was at times a lucrative living, even after the fees of literary agents. Again, first from his base in Rome in Via Poggio Catino, and then from Dortmund, the conversation between writer and editors continued. Some notes suggest that the need for each other was mutual. One memo asked: 'Is Bloke Modisane going off form?' 'Either off form or choosing the wrong material,' was the laconic reply. Then a more direct call went out: 'I know what an excellent writer you are. Why don't you send me an idea or two for an original Saturday Night Theatre which we could commission?' The note was from Shaun McLaughlin, it was August 1974, and the successful play *Leave Well Alone* was the eventual result.[36]

Modisane wrote his last (BBC-archived) letter to his closest BBC colleague, William Ash, at Broadcasting House in October 1978. As in his letters to Langston Hughes, he was able to sound desperate, funny, and serious in quick succession: 'If this goes on I am going to have to join the Rolling Stones. I have changed countries three times in five years and have now changed address again but this time in the same country.' Yet he was also able to tell Ash in the same letter that he had done work on *Sitting Duck* 'along the lines suggested by you' and to acknowledge Ash's 'invaluable advice with my plays over the years'.[37]

As Ash moved from the BBC, others within the script unit were there to continue the work with Modisane on his script for *Sitting Duck*. Bernard Krichefski was to work most closely with him as it moved, again, slowly, towards the kind of perfection that the script unit wanted. Even after the formal commission of a 90-minute play was tied up (the payment now £1,000),[38] the shaping was not complete and Krichefski could write warmly: 'I look forward to receiving the script in due course. Meanwhile if I can be of any help at any stage, please do let me know.' Some months later, after Modisane had made a visit to London from his Dortmund base, a note within the unit could say: 'Saw author ... [script] vastly better.'[39]

Sitting Duck was broadcast on Radio 4 as the Monday Play on 11 August 1980. Its producer, Brian Wright, saw it primarily as a racy, ironic commentary on war and its futilities, a 'stylized comedy'.[40] It was certainly comic, with strong echoes of the 'Good Soldier Švejk' figure in the novel by Jaroslav Hašek (2005 [1930]): its central figure a small but immensely resourceful man, caught up in empty battles in the endless

[36] Ibid. *Leave Well Alone* was rehearsed on 4 and 5 March 1976, edited on 9 March, and transmitted on Radio 4 from 3.05 p.m. to 3.55 p.m. (BBCWA Script Library).
[37] Modisane to Ash correspondence, BBCWA, Modisane Scriptwriter File 2, 1973–79.
[38] BBCWA, Modisane Scriptwriter File 2 1973–79 (3 January 1979; 14 February 1979).
[39] BBCWA, Modisane Scriptwriter File 2, 1973–79, Clare Taylor to Bernard Krichefski.
[40] BBCWA, Modisane Scriptwriter File 2, 1973–79, Wright correspondence.

tugs of power and cynical manoeuvrings by distant brokers. Yet the play also set out starkly the plight of civilians in wartime, in this case in the border wars between the South African army and the guerrilla forces of the liberation movements.

War, comedy, irony

As the pressure on the apartheid regime increased, and the end of apartheid began to seem a possibility, the general coverage of South African politics by the BBC, on both radio and television, grew substantially. One particular site of interest in the 1980s was the South West Africa/South African border war, with the forces of the South West African People's Organisation (SWAPO) and Frelimo pitted against the South African Defence Force (SADF). The early 1980s was a period of intense SWAPO activity; there was terrific South African military pressure on the front-line states and the MK cadres of the ANC were active in Angola/South West Africa. The intimacy of radio, its 'blind' aurality, meant that the small tragedies of such a war could be brought poignantly into the listening space. Could Modisane's play also have been part of a growing globalised awareness of the South African apartheid state's predatory and defensive cross-border wars as it tried to defend its policies on race?

On 29 January 1981, on the prime-time World at One on Radio 4, hosted by Robin Day, a former British mercenary, Trevor Edwards, gave a chilling account of the fighting tactics of the South African forces in Angola and what would become Namibia. He described coming to an abandoned SWAPO village; how the soldiers did a sweep of 50 kilometres and all civilians encountered were killed. Children under a bush rose up to run as the soldiers came and were shot; a man began to run and was shot; male adults were stripped and shot with their hands above their heads. He described how the mother of a shot child and the wife of a shot man followed the troops all morning, silently.[41] *Sitting Duck*, for all its intentional absurdities and its branding by the producer as a 'stylised comedy', captured such tragic histories and the plight of civilians caught up in such a war.

Set 'somewhere south of the Zambezi', *Sitting Duck* engaged in a powerful and direct way with the war of attrition that South Africa was waging on its northern neighbours by 1980. It had as its central voice the Švejk-like figure of Sergeant Damara, who wished to change sides, to join the guerrilla fighters and leave the SADF. But no one would trust him, nor believe his motives. The play made great use of the ability of radio to

[41] I was listening to the programme and made notes.

move quickly in time and space and to splice through carefully chosen musical bridges. Besides examining the plight of the civilian population, it explored the dilemma of a man who wished to change sides but was, perhaps, too implicated in the dirt of the side he wished to leave to be able to do so without the utmost difficulty.

As in past productions the cast list showed the black acting capital on which the BBC could draw. Sergeant Damara was played by the South African Alton Khumalo. Lionel Ngakane, who had played the gangster Pontiac Mick in *The Quarter Million Boys* a decade earlier, now doubled as the leading village elder, caught between two destructive forces, and the vacillating priest, who in the end could not help Sergeant Damara. Two decades earlier Modisane had swapped news about the musician Todd Matshikiza with Langston Hughes.[42] Now his son, John Matshikiza, at the time a member of the Royal Shakespeare Company, played the role of the enigmatic MK leader Bra Namib. Louis Mahoney, the Gambian-born actor who was now a part of the black theatre community in Britain, was also in the cast.

Sitting Duck, with its indictment of the SADF and its cautious endorsement of the fighting forces of the liberation movement and MK, signalled an acknowledgement by its author that the armed struggle was necessary. However, the play was full of ambiguities: there was no heroic figure and no valorisation of war. The short episodes and racy music, the agony of Sergeant Damara as he decides to change sides and move to MK but is refused a home in the new force, all suggested the exploration of ambiguous loyalties, painfully worked through. Eventually, Bra Namib accepts the deserter, and, in a flashback, their association in a distant time in a township in South Africa is briefly evoked. Ambition and idealism sit close together. Damara, who narrowly escapes being shot by both sides, is instead sent to Lusaka for 'training', to return as a general in MK, but only a two-star one, and not a five-star general as he had hoped.[43]

There was one play to come. *The Somebody* was broadcast on 13 July 1987 at 8.15 p.m. as the Monday Play, on Radio 3. Modisane had died in Dortmund in February of the previous year. Never one to abandon ambiguities, as his own life had contained so many, Modisane set a Mandela-like figure, but a flawed hero, at the centre of this play, which was set in South Africa in a place he called 'New Sophiatown'.[44] With all the ingredients of a dramatic political thriller, the play was full of the imagined sounds of an almost free South Africa, the excited voices of the people, popular cries of disaffection, and the plotting dialogues of those

[42] NELM, Hughes/Modisane correspondence, June 1962. [43] *Sitting Duck*, p. 77.
[44] I had no access to the BBC records beyond 1979. There may have been much give and take in the final production of this play too.

vying for power. It also included – more so than any of his other radio plays – the interior dialogues and private faces of its public characters.

The Garveyite character of Kalamazoo, longing for glory, sits with the depressed old leader, Kei Amatola. 'I'm an invisible man! To be famous is what counts,' he tells him. There is, too, the calculating politician of the new generation, Duke Marachudede, skilled in struggle rhetoric and quoting Raymond Chandler. In an unlikely scenario totally without political correctness but with many warnings for the future, Kei Amatola is assassinated in a set-up and Kalamazoo is thrown to the lions in the wild north by the leaders of the struggle. A leader of the old regime attempts to rescue him by helicopter as lions roar. 'Don't forget to sleep up a tree tonight,' warns the voice in the thick Afrikaner accent of the old regime as the play ends.[45] Dense, too, with discussion of past black history in the form of dialogue, and constantly throwing up the empty promises of earlier regimes as a warning for the future, the play was a challenge to any utopian vision of a free South Africa.

This last drama also marked Modisane's mastery of a genre and a medium he had lived with for almost 30 years. It was one from which he had sometimes sought to escape as the romance of film beckoned, but he was a master of the plays and play of sound. With or without the team of fastidious gatekeepers/aficionados in the BBC script unit, it was he who 'made' the plays. He was always the artist outside, but in a broader sense he was also 'within'. As the *Radio Times* of 13 July 1987 advertised its Radio 4 Monday Play, it also carried a write-up on *The Somebody* and its late author, William Bloke Modisane. Joseph Marcell, the Afro-Caribbean actor who played the Mandela figure of Kei Amatola, could say of him: 'Not only has he written a drama very much of our time, he was someone who made a huge contribution to British theatre.'

Accessing the radio works – and Modisane's involvement in a particularly fluid period of the BBC – allows us to better understand how a structure of power such as the BBC, which was both imperial and postcolonial in its impulses, was able, for a while, to intersect with and articulate the energies and creativity of exiled and diasporic artists from Africa and the Caribbean. The radio plays, and 'the world outside' from which they came, also allow a better understanding of that difficult space of exile from which Modisane was forced to operate. The ruptures of time, the splicings, the collages that he handled with such brilliance in *Blame Me on History* (1963 [1986]; Sanders 2002) were all devices used with special power in the aural medium of radio. Moreover, the nightmare of the modern, and in his case the black modern and the creation of

[45] *The Somebody*, p. 73.

a black modernist textual radicalism (Williams 1989: 35), could present itself with peculiar efficacy in this 'blind' medium of the voice, with its magical, Ariel-like qualities. Through the dramas in particular, Modisane was able to convey his preoccupations about modern South African society and the role of the intellectual, and he could maintain an intense engagement as an artist with the unfolding struggle in the Southern subcontinent during the late 1970s and 1980s. But there was more than that: his work had weight; it was respected and admired by the fastidious team at the BBC that was responsible for making radio drama an art form to be reckoned with in an era when television was king.

The medium of radio was never one that Modisane either wished to or could have had a place in within South Africa; even in the 1950s, there was no place for him as an urban black intellectual – working through the medium of English – on its heavily monitored airwaves (Gunner 2000a). Yet through the voices he created in the transient yet intimate and far-reaching medium of the radio dramas he wrote for the BBC in his years in exile, he could construct possible communities in Africa, set them up, and then see them move back into silence as the cast lists were read out and the daily programme on what was at first the BBC Home Service and then later Radio 3 and Radio 4 rolled on. Always, though, there was the voice of the uncomfortable, the outsider. But as the BBC itself shifted, he, too, through his long association with script readers and producers, was a member of a new community of the air, in a space that could command a new plural public and audience. Certainly, Bloke Modisane's dramas – and his wider 'voicings' had many publics (Warner 2002: 49–90). They spoke directly to the plural British public when they were heard as the Monday Play on Radio 3 or Radio 4; and they were beamed to a vastly different plural public across anglophone West and East Africa when broadcast in the African Theatre section of the World Service from Bush House. Possibly, the radio plays, honed in meshes of often difficult and sometimes dismissive dialogue with the editors 'within', enabled him to claim a rhetorical space for himself as someone who sometimes embraced an African nationalist position and yet simultaneously questioned any simplistic unitary nationalism. Possibly, too, they marked and changed the BBC itself.

4 'Africa on the rise'

The early 1960s, and the radio voice of Lewis Nkosi

> Africa will be free before we have even managed to get ourselves a lousy cup of coffee.
> James Baldwin[1]

> I see the future of the Negro anywhere in the world as indissolubly linked with the future of Africa.
> Lewis Nkosi[2]

Diasporic radio man

No one represents the fragmented nature of South African letters better than Lewis Nkosi. The focus of this chapter is on yet another 'fragment' of Nkosi's life, a hidden section of his radio work in early 1960s London. Central in the narrative is the idea of 'instability' and the creatively destabilising role that can be attributed to certain forms of radio in a particular era (Brennan 2010). I consider in the chapter Lewis Nkosi's involvement in the early 1960s with the Transcription Centre, the maverick outfit set up by Dennis Duerden in 1962 in London that ran for seven years, funded first by the seemingly innocuous Congress for Cultural Freedom, then by the Farfield Foundation, and finally closed for lack of funds in 1969.[3]

Nkosi's diasporic energy, his understanding of art, power, and global black cultural connection, flourished in the racially tense, vibrant milieu of 1960s London (Farred 1996). It was a time of new ideas on the transference of culture, when debates on cultural mutuality rather than one-way dominance were on the agenda. The early years of that decade saw Lewis Nkosi living in London after his time in North America, publishing essays in a range of journals, literary and political, and finding a niche in radio, still an important medium in Britain and one of growing importance in Africa. It was also a medium clearly recognised as powerful by the South African apartheid government, which had driven Nkosi into exile with a one-way travel permit a few years earlier. The offices of *Drum* and *Golden City Post* in Troy Street, downtown Johannesburg, were

1 James Baldwin quoted in Modisane radio interview, c.1963, Transcription Centre Africa Abroad programme, London.
2 Nkosi, preface to *Home and Exile* (1965).
3 Harry Ransom Humanities Research Center (HRHRC), Transcription Centre (TC).

Figure 4.1 Before London and before the US, Lewis Nkosi, print journalist and later media man, at his desk in the *Drum* offices, Troy Street, Johannesburg.
Source: *Drum* photographer, BAHA/AMO.

a stone's throw from the black radio workers' entrance to the SABC, and close to Park Station – that hub of daily arrivals and secret departures. While black radio artists such as Alexius Buthelezi, Winnie Mahlangu, and others arrived in Troy Street for their broadcasting, which would build up a station with huge popular appeal among African-language listeners, Lewis Nkosi took another radio route in the streets of London.

Why should it be important to focus on a tiny part of Nkosi's creative life, connected to a small and, many would say, failed broadcast and cultural outfit, the Transcription Centre?[4] It was a venture set apart from

[4] The tapes from which I worked for this chapter and for Chapter 3 are part of the Archival Sound Recordings now held at the British Library, London, under the heading of 'African Writers Club'. I searched only under 'Nkosi' and 'Modisane'; the whole collection is far larger. The title 'African Writers Club' itself is confusing as many of the recordings are

the great national media institution of the BBC, which sailed on magnificently, broadcasting and televising to the nation from Broadcasting House, Portland Place. In addition, it did not forget the empire: it relayed programmes to listeners in African countries through the African Service at BBC Bush House on the Strand.

Nkosi was known and appreciated by the energetic producers at Bush House, but his voice and role in the Africa Abroad magazine programme, which Dennis Duerden initiated as soon as he set up the Transcription Centre in January 1962, made their own distinctive mark. The space gave Nkosi the opportunity to capitalise on his unique set of diasporic and cosmopolitan skills, honed first in the 'global ecumene' of Sophiatown (Hannerz 1997) and then during his period as a Nieman Fellow at Harvard. The impact of this time in North America on Nkosi's sensibility with regard to race and history is captured in his essays on jazz in *Home and Exile* (1965). His shocked encounter with life and conditions in the American south permeates most of the book and is a key to understanding his self-positioning vis-à-vis the African continent and African-American 'Negro' links during this crucial period of the early 1960s. Nkosi had a free space in his Africa Abroad programmes in which he could exploit his skill in reading the fluid landscape of history and blackness at that moment in imperial and postcolonial history. He could be a diasporic radio man, an interpreter of subtle moves of history and culture, and a translator of blackness in a capital itself in a state of flux in late empire.

A key element in Nkosi's ability to operate with moments of freedom not possible in the more controlling atmosphere of BBC Bush House and the Africa Service was the nature of the Transcription Centre itself and the outlook of its founder, Dennis Duerden. An article in the magazine

from the Transcription Centre, and are specifically from the Africa Abroad programme recorded at the Transcription Centre. This is quite clearly stated in the recordings, which seem at pains to state their origin, as a kind of copyright marker. At times, though, there are 'African Writers Club' programmes, such as that of Edward Blishen interviewing Bloke Modisane after the publication of *Blame Me on History* (see Chapter 3). This would have been a BBC programme run from Bush House on the Strand, London, not one from the Transcription Centre. Or so I assume. How much movement of programmes and recordings took place between Bush House and the Transcription Centre is something that still has to be carefully tracked. I have not seen all the relevant boxes at the HRHRC, University of Texas. Many of the recordings held at the British Library in their now digitised Sound Archives Collection are catalogue numbered but not dated. When they are dated, the date may be wrong. Voices that are recognised are listed, but others are not. In the important recording where Lewis Nkosi interviews Bloke Modisane after the latter's lecture tour of Negro colleges in the American South, Modisane's voice has not been recognised and the catalogue note states merely 'unidentified male speaker'. All recordings from the Sound Archives, British Library, are marked as 'AWC' (i.e. African Writers Club), although many are actually Africa Abroad programmes from the Transcription Centre. The links between the important broadsheets, 'Cultural Events in Africa' (now housed at the Centre for African Literary Studies (CALS), University of KwaZulu-Natal, Pietermaritzburg Campus; formerly the Bernth Lindfors Collection, University of Austin, Texas), and the smaller collection in the HRHRC at Austin, and the recordings from which I draw for Chapters 3 and 4 have still to be established.

West Africa,[5] written almost five years after the opening of the centre, gives some sense of its essential difference, and of the nature of the man himself. The writer of the article, 'a correspondent', describes the Transcription Centre as 'a kind of African cultural oasis in the London desert' and ascribes most of the credit for this to Duerden.

What is not overtly stated, but is implied, is that Duerden's idea of a certain permeability in the nature of African cultural forms was part of what drove his particular vision for the Transcription Centre. His own creative life as a painter, his work with young artists in Keffi, northern Nigeria, and his time as assistant curator at Jos Museum before his return to England was followed by a spell at BBC Bush House, where, in 1958, he became a producer in the Hausa Service. Clearly, though, Duerden yearned for a free space for the production of a pan-African or diasporic African voice, something that he could not find in the Bush House Africa Service. Possibly, it was the subtle constraints of the voice of empire, summed up in the word 'ponderousness' (see below), which he longed to escape (Bloom 2014). In a letter to his last funder, Frank Platt of the Farfield Foundation, not long before the Transcription Centre closed shop, Duerden gave a hint of why he had struck out on his own, and why, even when the Transcription Centre seemed too much for him, he had found the idea of returning to the BBC African Service difficult to contemplate:

> I found it extremely difficult to work in large organisations because of the ponderousness of the men that I had to deal with. About three years ago I played with the idea of going back into the BBC and was informed by the Head of the BBC African Service over lunch that the Service would find it impossible to accommodate working with a man such as Wole Soyinka because his temperamental idiosyncrasies would be disruptive and would prevent the smooth working of the Service.[6]

It was thus a search for the possibility of debates running unhindered between the creative minds of the era that partly drove Duerden to find a space outside the at times oppressive institutions of the BBC. In the same letter to Platt, alluding to the mandarins of the BBC, he wrote: 'I am far too impulsive and impatient a character to spend hours discussing my projects with pretentious beribboned gentlemen.'[7] It was the constrictions of the BBC African Service that he longed to escape. Moreover, the sheer efficiency of BBC operations seemed to restrict an expansiveness of African debate, and it was this that he hoped to bring about through a free, even promiscuous, capturing of the aural and auditory through the medium of radio.

Duerden spelled this out in his letter to Platt, as he recalled precisely the moment of birth of the idea for what became the Transcription Centre:

[5] Dennis Duerden, *West Africa*, 28 January 1967, p. 101.
[6] Dennis Duerden to Frank Platt, 18 June 1969, HRHRC TC boxes. [7] Ibid.

> In fact the TC started on Saturday afternoons when I was producing a series of programmes for the BBC called 'The Future of African Society'. These were extended discussions which finally had to be butchered to make up 15 minutes as that was the only time available for broadcasting.
>
> We recorded from 30 to 60 minutes of discussion and spent two or three hours afterwards going on talking in the Canteen, which on a Saturday afternoon was deserted . . . It was as a result of my producing these programmes and meeting in connection with them people like Ezekiel Mphahlele, Okot p'Bitek, Godrey Leonhardt [sic], Alan Ogot, David Rubadiri and Donatus Nwoga that the TC was first started.[8]

As the article in *West Africa* put it, rather more politely, Duerden felt that there was a need for 'a new kind of radio service in which there would be greater opportunity for contact and interchange of ideas between artists, writers, musicians, actors and critics than is usually afforded by the large broadcasting organisations'.[9] Although the new centre had awkward relations with the BBC African Service (it was seen as a competitor, the article hints), it worked well with other sections of the BBC, and in 1967 Duerden advised the Third Programme on its production of Wole Soyinka's *The Lion and the Jewel*. Certainly, too, the Transcription Centre rapidly built up 'a substantial clientele for its tapes, not only with government radio stations in English-speaking Africa but with radio stations . . . such as the encouragingly progressive Deutsche Welle (Voice of Germany)'.[10] Its reach, in fact, was even wider than that:

> By December 1965, 516 of its tapes had been broadcast by radio stations in 21 countries, including 9 stations in African countries . . . as well as in the USA, UK, Sweden, Federal Republic of Germany, India, Malaysia etc.[11]

It was, though, always an uncertain venture, constantly needing funds for rent and patronage, with Duerden moving with a certain creative restlessness from one enterprise to another, each within the ambit of 'African culture'.

London, the Transcription Centre, and 'the Nkosi period'

As the main voice and planner of the Transcription Centre's Africa Abroad programme in the opening years of the centre, 1962 and 1963, Nkosi's role as global interpreter and path-breaker was crucial. Away from the 'butchering' constraints of time laid down by the BBC African Service – as expressed by Duerden – there was a place in the Africa Abroad programme for expansiveness, value, and weight on debates on race, and 'Africa', if skilfully handled.

The Transcription Centre, operating its 'cultural oasis' from rented premises in west and later central London, was itself in a city in flux. The

[8] Ibid. [9] Dennis Duerden, *West Africa*, 28 January 1967, p. 101. [10] Ibid.
[11] HRHRC, notes on the Transcription Centre archive.

Notting Hill riots of 1958, just three years earlier, had seen the politics of race establishing itself as a crucial factor in the life of the metropolis (Fryer 1984: 378–9, 381). As Britain's white majority demonised the black presence, so too did a more militant awareness of Britain's institutionalised racism come into being (Hall 1996: 245, cited in Farred 1996: 41). It was this 'second' generation of Afro-Caribbean immigrants, more prosperous and often more educated than the earlier newcomers, who drew on the inspiration of African-American models of resistance and moved to a more activist black consciousness position which was part of the black presence in London.[12] The British capital in the 1960s was marked not by the brilliant swirl of 'black Paris' in the 1920s and 1930s (Edwards 2003), but by the presence of its own new wave of Caribbean migrants, as well as the glittering diasporic strangers who passed through (James 2003 [1932]; McLeod 2004). The Caribbean migrants' lives are vividly and poignantly evoked in Sam Selvon's novel *The Lonely Londoners* (1956) and in George Lamming's ironic and autobiographic *The Pleasures of Exile* (1960). There were others besides Selvon and Lamming writing and working in London. Andrew Salkey's edited *West Indian Stories*, published by Faber in 1960, marked the recognition of Caribbean voices in London's literary world, and V. S. Naipaul's picture of a decaying London, unsure of its nascent multicultural identity, is captured in the sadly comical *Mr Stone and the Knights Companion*, first published in 1963 (1988 [1963]).

Parts of London, though, had long reflected the cosmopolitan flavour of the city, which was to become more marked from the 1960s. Soho, slightly to the south of Portland Place, where the BBC's main centre, Broadcasting House, stood, and north-west of Bush House on the Strand, from which the World Service (and African) programmes were broadcast, was one of those spaces. In 1935, Paul Cohen-Portheim's *Spirit of London* (cited in Eade 2000: 54) described Soho's outer confines as 'a picture of lively exoticism' filled with Greeks, Armenians, Cypriots, and Negroes. John Eade comments on the change in Soho a few decades later:

> From the 'Swinging Sixties' onwards, Soho's image subtly altered as the apparent certainties of nationalism and imperialism gave way to a more willing acceptance, however uneasy and contradictory, of multicultural diversity and global influence. (Eade 2000: 65)

The writers, politicians, and intellectuals who were briefly actors and fellow conspirators in the African and transatlantic flow of ideas and debates of the early 1960s thus entered a London full of fluidity and change. It was a hub for Africans passing through, meeting, talking and being caught 'on air', sometimes by the BBC African Service, but

[12] Nkosi's 1969 BBC radio play *We Can't All Be Martin Luther King* dwells brilliantly on this.

from early 1962 also in the far wider, baggier net that the Transcription Centre provided in its Africa Abroad programme, with Nkosi initially presiding. The visitors' voices on air gave a sense of their passages across oceans and continents, like migrating birds in flight. The criss-crossings of their journeys – between African countries, North America, the Caribbean, Britain, and sometimes France – suggested a new sense of agency and purpose, and the birth of a new post-imperial era. The Transcription Centre debates and interviews that Nkosi conducted with francophone writers such as U'Tamsi and Mbelle Sonne Dipoko – although in English – made it clear that a unity of specifically African interest cut through the boundaries created by the former colonial powers of France and Britain. London, in its cultural life and its self-projection as a city, was moving between a kind of late imperial cosmopolitanism and a reactive British nativism. The sheer nimbleness of the Transcription Centre radio presence, operating like a clever tiny animal so common in African folklore, allowed it a manoeuvrability of discourse that may have eluded the more ponderous, elephant-like BBC.

Nkosi himself embodied and voiced this nimble presence, shaping and mediating, as he took up his work on the Africa Abroad programme in the Transcription Centre studio, a movable and relatively impromptu space in a fluid and volatile era. He was operating not simply as a South African writer but as part of something much wider, within a broader sphere of thought that spanned the political and the artistic and related to the politics and art of the African continent, and of black America and the Caribbean. The late 1950s and the cusp of the 1960s were key moments in the growth of transnational black consciousness and the rise of Africa in world affairs, as the continent not only began to throw off the thrall of colonialism, but also became a player in the politics of the Cold War (Baldwin 1963: 27; Borstelmann 2001; McDonald 2009). By the end of 1962, Ghana and Nigeria were already independent, as were Algeria and Uganda. Kenya was to follow in 1963, and Malawi and Zambia in 1964. The Caribbean, too, was on the move, with Jamaica, Trinidad and Tobago all gaining independence in 1962. The former Belgian Congo had received its independence in 1960, and Patrice Lumumba had been assassinated under murky circumstances in 1961. In North America, the civil rights movement was gaining ground, as the Freedom Rides of 1961 and the deeply contested, although finally successful, entrance of James Meredith to the University of Mississippi in 1962 showed (Dudziak 2000: 157–64). But a victory of equal rights for all American citizens seemed a long way off.

Radio, connections, and counter-discourse

As Nkosi took up his new brief in January 1962, the perception of intense links between the fate of Africa and of 'the Negro' surfaced often in discussions. There was a scent in the air of something new, and the radio voices from these programmes captured this well.[13] The interaction of speakers and their existential, free-flowing discussions (which can still be heard, with the original tapes now digitised)[14] suggest the intensity of the debates among artists, thinkers, and people on the ground – the weight of their voices operating as counter-discourse within a larger flow of historical domination. Was this a moment of possible reconfiguration, when Africa, perceived as a new force in the world, politically, artistically, and spiritually, appeared to speak to the transatlantic black world, to itself, and to its former colonisers, in tones of confidence and authority that called into question and in some ways overrode the sufferings of the past?

Radio allowed a mixing and merging of voices and ideas in a kind of flow of presentness, which, if listened to over half a century later, gives an impression of the uncertain and urgent pushing to and fro of ideas and positions on culture, identity, and politics, and on the question of African and black American identities on the global stage. At the time, as these conversations travelled to their various audiences across Africa and Europe, they must have provided an instant and acute sense of a continent on the move, of 'Africa on the rise'.

For Nkosi, power as politics and the discursive power held in African culture in its modern forms could not be separated. Symbolic and cultural capital were both part of this moment of emergent political power. The programmes returned to this again and again as figures such as the Nigerian poet Christopher Okigbo, playwright Wole Soyinka, Ghanaian philosopher Willie Abraham, and poet and essayist Langston Hughes interacted. Intensely, at times humorously, they placed Africa in the world along with literature, music, art, film, theatre, philosophy, and Africa's future (see Viljoen 2007). It was, though, largely a male discourse. Women featured hardly at all, although Nkosi would refer with gusto to 'the voice of Mother Africa'. The Africa Abroad programme projected the voices and views, the doubts and hopes of speakers to an audience of the ear that spanned continents. In its diverse and often unexpected confluences of genres, places, and voices, the programme mediated its cargo of new symbolic capital in a medium that was to become Africa's own, that of radio. This was a new genre for a new, expansive moment in radio and in black cultural and political negotiation.

It was, crucially, a technology of voice, as the crisp tones and accents from Africa, the Caribbean, and North America rubbed up against those of a more recognisable cultivated Britishness. The very mix and unplaceability of its 'pieces' of Englishness suggested things strange yet familiar. The 'estrangement' of English sound caught the beginnings of something

[13] From listening to original, now digitised, tapes at the British Library. See note 3 above.
[14] See note 4 above.

new – language in search of new paradigms of self, nation, region. Soyinka's and Nkosi's voices – the former quite upper-class British in its intonation but with sudden wild incursions of another tonal system; the latter slower, drawling, mimicking British-received pronunciation – bumped up against English spoken in francophone accents from the Congo, the voices of Tchicaya U'Tamsi and Mbelle Sonne Dipoko, or of Bernard Fonlon from Cameroon. 'Negro' voices from North America, with inflections of the south or of New York, voices from Kenya and from Tanzania, all had their place on the Africa Abroad programmes anchored by Nkosi in the Transcription Centre's first two years. In this transcontinental and diasporic exchange of talk, music, poetry, and song, the languages of Africa sometimes crossed tantalisingly with the old imperial language. All was mediated by the modern and modernising technology of radio.

With London as a hub and the Transcription Centre as a hive of radio connectivity, Nkosi was able to capture an assortment of views on culture and politics in a new post-imperial moment as he made use of fleeting visitors. African sojourners in America were pressed by him on topics such as 'the new mood among Afro-American people' and 'how young American Negroes are deciding to free themselves'. The reply, from a francophone visitor, showed the necessary 'translations' between divergent black cultures at work (Edwards 2003: 2–3). 'As an African Negro I was very shocked by lots of things I saw after a year in the USA, having come from the Sorbonne,' was the comment of a Mr Thiam, speaking on the Africa Abroad programme.[15]

On African politics, Nkosi mediated divergent views, conjuring up impressions of new public spheres where novel kinds of political debate could flourish. In one issue of the programme, speakers discussed the Constitutional Conference under way in London, where the blueprint for an independent Kenyan government was being thrashed out. Ronald Ngala spoke for the Kenyan African Democratic Union (KADU), arguing for a more federal structure. Tom Mboya, for the Kenyan African National Union (KANU), argued for a more centralised one.

The radio debates also morphed into the print ephemera of the time, as journals, too, attempted to capture the new debates and intrusive ideas. The Transcription Centre received a number of key African cultural journals, and early in its short life began the newsletter 'Cultural Events in Africa'.[16] This ran as another strand of its network of recordings, and, in 1960, had 236 subscribers. Nkosi saw journals, too, as vectors of new identities and as cultural 'translators' (Edwards 2003: 7–8) in the dialogue of divergent black heritages. *Black Orpheus*, from the Mbari Centre

[15] C134/422, National Sound Archives, British Library. [16] HRHRC TC.

at Ibadan, and subsidised by Longman Nigeria, was regularly sent to the Transcription Centre. For Nkosi, the magazine *Black Orpheus* was important not only for the African and African-American work that it carried, but because it was a symbol of black creative power moving into a new conceptual space beyond the Greek myth of Orpheus and Eurydice. The Africa Abroad programme regularly commented on issues that featured in *Black Orpheus* and in the East African *Transition*, and Nkosi praised and critiqued as he saw fit.

A ploy that Nkosi often put to good use was to pose cultural questions and political imperatives in tandem, as if to emphasise that power resided in a complicated knot of the political and cultural. Arts and book review segments within the Africa Abroad programme regularly became points of departure for free-ranging discussions. Nkosi would often return to the big question: 'Is there a brotherhood of blackness?' Typically, the panel on 'Is there a brotherhood of blackness' consisted of the voices of men never women, and showed Nkosi preoccupied with the overriding questions of those years: 'Are we black men related? What is our certain quality? Is there such a thing as our distinctive outlook on life? Does it leap across different histories?' The panellists – the playwright Barry Reckord from Jamaica, Charles Patterson from North America, the Ghanaian philosopher Willie Abraham, and the Guyanese Alan Gibbons – came at the questions from a range of perspectives. Reckord saw class as the divider, not race. Nkosi, as host, trod a balance between intervention, abstention, and domination. At one point he quoted from James Baldwin's essay on the Paris conference of 'Negro-African Writers and Artists' (held at the Sorbonne in September 1956):

> What held [black men] together was their precarious and unutterably painful relation to the white world. What they held in common was the necessity to ... no longer be controlled by the vision of the world and of themselves held by other people. (Baldwin 1960: 35)

In the same programme, Gibbons, from British Guiana, suggested that the question of slavery was a basic source of unity. In the summing up, the philosopher Willie Abraham and the anchor Nkosi appeared – on this occasion – to agree. For both it was the experience of history more than the links of culture that bound them (black men) together.

Duerden's cultural entrepreneurship and the freedom he had with the Transcription Centre to operate outside the large media institution of the BBC meant that he could make his own deals. One was struck with the American National Educational Television (NET) network. This was for a series of interviews with key figures in 'the new wave' of African culture, and they would take place in 1963 in various African venues. Nkosi was the interrogator and facilitator; as the ideas man, the unpredictable *anansi* figure 'on loan' from the Transcription Centre, he presented and interviewed for the American NET series on 'African Writers of Today'. These programmes were shown on the network in the US in 1963, and extracts were featured in the July 1964 issue of *Africa Report*.[17] Nkosi, looking impish but serious, and holding a huge cigar with studied casualness, adorned the front cover of the issue. Each of the African writers Nkosi interviewed was also represented by a photographic portrait, as if to give weight to their words. The sites chosen for their discussions emphasised the weight to be given to their discourse, and were therefore usually a university library or a museum. All this gave the impression of an archive of power to be savoured in the present and stored for the future.

Because of the deal Duerden had struck with the NET, the recorded interviews were also available for the Africa Abroad radio programmes in 1963. Here, on the airwaves, what could be captured was only words and the flow of ideas, and the accents and intonations of the speakers, along with their nascent or already established reputations. Yet this itself was important. For, on radio, they were within the flow of sound mediatised by voice, and were thus entering Africa's oldest and still vibrant medium of communication: orality, with its huge store of genres, some still influential in the ethical, social, and political life of the continent (Gunner 2004). The oral space as a marketplace of ideas and an experiential and somatic archive was a structure with deep roots in African epistemologies, and radio as a medium for 'talk' was flowing into that element of powerful, weighty speech. But there was one important difference – in a sense, the speech in this series was never directly from the marketplace. It was the mediated passionate, articulate, and at times verbose expression of the new elite, not the raucous calls, songs, and performances of the marketplace. So, in the radio programmes, what mattered was the sound of the voices and the ideas, rather than the site of recording, as the talk seamlessly fed into the listeners' lives. Thus, on one Africa Abroad programme, Chinua Achebe and Soyinka discussed their work with Nkosi in the Nigerian National Museum in Lagos.

As part of the same series, Nkosi, along with Soyinka, debated the nexus of culture, politics, and the face of modern Africa with Willie Abraham, recent author of *The Mind of Africa* (1962), in the library of the University of Legon in Accra. On this occasion Nkosi was distinctly irritated by the idea of the archaeology of the past for its own sake. It was the modernising life of the cities that mattered most, he argued, and this was what should be captured by African writers. Abraham, however, disagreed. He argued the case for seeking the force of literary production in what he termed 'the classical oral tradition' in which 80 per cent of the

[17] *Africa Report* 9 (7): 7–21; HRHRC TC boxes.

continent still operated. And he reminded Nkosi that elements of these 'traditions' were active in the cities too. In these debates, which moved over the mediums of print, film, and radio, Nkosi played his role of cultural broker and mischief-maker. He was showcasing black intellectuals seeking their own destinies and setting out their own 'vision of the world', as Baldwin (1960: 35) had put it when he interpreted the Paris gathering of Negro-African writers and artists of 1956. Nkosi was a trickster figure and seeker, beating about for the free space outside the dominant discourse 'and disciplinary traditions in the main fields of ... social and cultural enquiry' (Said 1993: 47).

The Africa Abroad programme, however, was diverse, and it was by no means confined to the elevated debates held in museum and library. The programme moved restlessly, creating its own 'marketplace' of sounds. In the Christmas 1962 edition of the programme – thus at the end of Nkosi's first year at the Transcription Centre – he presented 'highlights from the past year'. True, they were 'highlights', but the range of textualities, languages, genres, and political and philosophical discussions that surfaced might have struck a producer intent on a certain tidiness as extraordinarily diffuse. Yet they summed up the range of African and African-American voices present in the new debates and new discourses at that particular moment, without setting paradigms or parameters.

Most importantly, it set such voices alongside each other, as if their very proximity in recorded time spoke of something much bigger, a deeper if more hard to define connection. The short snatch of Gambian kora music by the master singer and player Banna Kanute (a famed exponent of the Sunjata epic) (Suso and Kanute 1999), sat alongside the Ghanaian poet Frank Parkes' 'Blind Steersman', a poem lamenting Africa's powerlessness in the world. Joseph Kariuki, the Kenyan poet just down from Oxford, read a poem with a similar theme called 'The Expendable Beast'. Extracts from recordings made at the African Writers Conference held at the University of Makerere, Kampala in June 1962, which had comprised a full programme earlier in the year, included the light, high voice of Christopher Okigbo, the learned and provocative Nigerian poet, steeped in the classical texts of Rome and Greece. Okigbo was reluctant to consider African literature as a category, preferring rather to think in universals. Langston Hughes, who had been a dominant presence at the Kampala 1962 conference, also featured. He spoke defending the philosophy of Negritude, and likened it to aspects of the 1920s Harlem Renaissance. Hughes' coupling of the two periods and impulses is important as it linked discourses of black resistance from two disparate historical moments:

> [Negritude] ... seems to be very similar to the feeling of American Negro writers in the 1920s, the period of the so-called Harlem Renaissance ... we quite

consciously were trying to bring forward and lift up the beauty of Negroness, what they call Negritude, but we had no phrase for it, we had no formula, we didn't make a school of it consciously. It seems to me that is what Senghor and the other poets are now trying to do in their way ... the beauty of Africa, the beauty of blackness.[18]

Another item in the 1962 Christmas broadcast showed discussion pushing genre boundaries as it commented on styles of performance. Listeners were reminded of an earlier programme in the year that had touched on Hughes' gospel musical, *Black Nativity*, which was 'packing in audiences in the [London] West End'. As a snippet of *Black Nativity* was re-played, listeners would again hear snatches of a hybrid, emergent genre, that of the 'gospel song-play'. It was one on which Hughes himself worked after attending the gospel temples in Harlem. At first called *Wasn't that a Mighty Day?*, Hughes changed the name to *Black Nativity* to honour what Arnold Rampersad called his 'faith in the importance of a black racial sense' (Hughes 2002b: 300). Hughes' notes (although not on the Christmas 1962 Africa Abroad programme) are worth quoting. They show how he worked, and how he knew that this new black genre was stretching the boundaries of accepted American art. 'Gospel singers are a new asset to American theatre,' he remarked.

In preparing my gospel song-play *Black Nativity* that is currently touring Europe, I visited a number of gospel temples in Harlem, usually arriving about ten at night when sessions were in full swing. I was never bored. Song and a sense of drama swirled around me. A mingling of ancient scripture and contemporary problems were projected with melodic intensity and rhythmic insistence. Every night I was drawn into the circle of oneness generated by the basic beat of the gospel tempo. (Hughes 2002a [1961]: 353)

Jazz, so important for the South African *Drum* writers and the musicians of the 1950s as a way into the global racial imaginary (Titlestad 2004), was one of the highlights of the 1962 Christmas programme. It had featured in many issues, often appearing alongside indigenous singing and instrumental music from another part of the continent. In one issue, 'Pondo Blues' by Eric Nomvete and his Big Five from East London (South Africa) was followed by a comment from the music critic Julian Beinart. Nkosi intervened to draw jazz into the debate on what a contemporary African tradition might be. Jazz such as 'Pondo Blues', he thought, entailed the exploitation – and modernisation – of an African tradition. On the Christmas edition, listeners also heard Cannonball Adderley's 'Damn Dirty Blues'. Cannonball was a jazz musician seen by critics such as Frank Kofsky (2001 [1970]: 74, 89–90) as part of the cultural and political ferment about exactly who owned jazz in North

[18] Hughes, C134/239/02, National Sound Archive, British Library.

America, as black musicians were sidelined by the wider music industry and fought back.[19] Jazz, which snaked its way through so many of the programmes, played the role of a cultural common denominator crossing between the continents of Africa and (North) America. It could cut through notions of bounded 'tradition' and simply move ideas from one cultural moment into another.

At times in the programmes, jazz captured the link between cultural identity and political voice. At one point, the Negro-American sociologist Tom Blair was interviewed by Nkosi. Blair, in his soft voice, which became loud and tense only at specific dramatic moments in the discussion, explicitly connected the fight of Negro artists affirming rights for the American Negro with the rise of Africa in world affairs.[20] Blair recalled the moment shortly after the assassination of Patrice Lumumba (17 January 1961) when:

> more than 200 American Negroes burst [the volume of his voice soars on this word] into the United Nations in the middle of a speech by Adlai Stevenson and they said that, 'We denounce America and Adlai Stevenson for even suggesting that we American Negroes are against Lumumba . . . We reaffirm our interest in Lumumba'.

Among the leaders of those present, Blair continued, were 'young, creative American Negro writers, jazz musicians'. Setting this in a wider context of Negro intellectuals and the fight for citizenship, Blair quoted Baldwin's written response to the event:

> We believe that the American Negro is the heart and part and parcel of America and we have the right, as any other citizen, like any other American citizen to express our point of view.

On this occasion, Nkosi responded again with a rephrasing of a Baldwin remark from the 1956 Paris conference:

> Yes, I remember that, James Baldwin saying that hereafter the Negro shall no longer be controlled by the White American's image of him and he said that had everything to do with the rise of Africa in world affairs . . . (Quoted from Baldwin 1960: 35)

James Baldwin, Africa, and that 'lousy cup of coffee'

Baldwin's writings, with key ideas rephrased or snatches quoted verbatim like jazz riffs, featured often in Africa Abroad, particularly in the opening two years when Nkosi presented the programme. After 1963, Duerden's

[19] Kofsky (2001 [1970]: 79) remarks of the track 'Damn Dirty Blues' that the title – which he has as 'Them Dirty [i.e. Funky] Blues' – 'that [it] is, a characteristically black way of playing the blues'.

[20] C134/26/3, National Sound Archives, British Library.

interests at the Transcription Centre moved away from his original aim of focusing on 'radio tapes' towards a more diverse African cultural presence in London. The radio presence lost its role as tricky African spider to the BBC elephant and gradually faded in significance. There is one moment in 1963 (apart from the 'African Writers Talking' interviews, which had a television, print, and radio presence in the Africa Abroad programme) when a capsule of sound in the digital archives captures the voice of Nkosi in conversation with what the records note as an 'unidentified male voice'. As soon as I heard the tape in the British Library Sound Archives, it was clear that 'the unknown African man', the second speaker in the conversation, was Bloke Modisane.

The importance of the transatlantic connection for these two men, both exiles from South Africa, and the burning question of the Negro-American freedom struggle dominate the exchange. There is no mention of the historic march on Washington and Martin Luther King's 'I have a dream' speech of 28 August 1963, so this discussion must have predated that. The two men tease out the meanings of the African–Negro-American axis, and this time it is Modisane who speaks with authority, drawing on his recent North American lecture tour. Unlike the discussion on the North American tour of K. E. Masinga only a few years earlier (see Chapter 1), the two speakers, interviewer and interviewee, compatriots in exile, tussle with ideas as equals. They are outside the unequal domain of power that held Masinga and his friendly, if somewhat remote, SABC English Service interviewer. Each is aware that they exist within larger global domains of inequality and it is this that guides their talk of African freedom and Negro rights, and their historical moment. Nkosi interrogates Modisane about his recent tour of America. Invited by the United Negro College Fund and the American Society of African Culture, Modisane had visited and lectured at ten Negro colleges in the south and one in New York.

Their discussions range over Negro literature, the 'genuineness' of Negro interest in Africa, the separateness of the communities in America, the feeling of space in New York and America, the inequalities of education in North America, power, and identities. Their opinions play off each other, as do their voices. Nkosi has his distinctive drawl. Modisane speaks tersely, quickly, and with a characteristic tension. In the extract here, they bring in Baldwin as a touchstone, and swing to Africa and to the humiliations faced by black Americans:

NKOSI: How genuine is this Negro interest in Africa? Is it an interest that is being worked up? Is he using it as a lever for his own freedom?

MODISANE: ... Now that Africa has come to be associated with power, Negroes feel that it is a good thing to be associated with Africa so there is a great leaning towards Africa. James Baldwin said, 'All of Africa will be free before we can get a lousy cup of coffee.' They have come to feel that their greatest

> hope is to help Africa in her fight for liberation and freedom. The more Africa becomes free the more they will become free. I don't think they want to be Africans but it is a power struggle, a struggle of shifting power.[21]

Modisane and Nkosi were both deeply invested in Negro-American politics of identity and culture and in the question of equality in America. Both admired Baldwin, and in the full radio interview from which the above is drawn, they refer to him several times as they tease out further the question of Negro and African identity, and wider American identity. Baldwin, for them, was a seer, a touchstone for their own thinking on race and justice in both South Africa and North America, and on black diasporic identity and transnationalism. The essays in Baldwin's *The Fire Next Time* (1963) and *Nobody Knows My Name* (1960) sit tight behind their musings. 'Will it be the end of Negro identity once oppression is lifted from his life?' is a question Nkosi puts to Modisane at a later point in the same interview, referring to Baldwin and 'one of the interviews done with him recently'. The answer again suggests how Baldwin's thought was etched in the minds of each speaker:

> MODISANE: James Baldwin also says that the salvation of White America is in the hands of the Negro.
>
> NKOSI: Yes, he is almost metaphysical in his analysis of the relation between the white American and the Negro American.[22]

Through the more open space of the Transcription Centre, the interview may have had a fullness that it would not have had in the more tightly managed BBC time slots – where long leisurely discussions on Africa from the literary thinkers of the day were, in Duerden's words, 'butchered to 15 minutes'. Finally, though, it could do nothing more than perform again the power of speech. It could demonstrate the give and take of different positions, show threads of influence running through conversations that were global as they laid out what was at stake: culture in a wider nexus of power, politics, and influence in the context of the Cold War, a restless black America, a nervous American administration, and an uncertain Africa.

'Transcription', translations, words, and place

The alternative space of Duerden's Transcription Centre, the BBC African Service, housed as part of the BBC World Service in Bush House on the Strand, and the great Portland Place building itself, home to the main BBC, each existed in the longer history of British Empire and race. Duerden's centre was, for a moment, an experiment in escape and

[21] AWC C134/454, National Sound Archives, British Library.
[22] C134/454, National Sound Archives, British Library.

recreation, carving out a space for more talk, and a freer capturing of voices and positions that were sent out to stations eager or curious to catch the new opinions and alternative positions on the cusp of huge global shifts – or so it seemed at the time. But the Transcription Centre was itself caught up in the toils of the politics of the Cold War, as the Congress for Cultural Freedom – unknown to Duerden – was linked to the American FBI. London in the early 1960s was without doubt a hub for black writers and intellectuals who performed new identities and forged new links, and radio played its part in this. It had, too, another identity as a site of race prejudice and fear of blacks, and thus a kind of double identity, liberatory and oppressive.[23] In this way, it resembled the Paris of the 1920s, where, as Edwards reminds us – it was sometimes easy for those of the African diaspora revelling in its freedoms to forget the underlying racism of French African rule in the colonies or to factor it into the present moment of black internationalism (2003: 4–6).

Thus, what I have called 'the Nkosi period' of London in the early 1960s had its roots in a much longer African and black presence in the British capital, and in other European capitals, primarily Paris. The fact that francophone thinkers and artists also had a place on the Africa Abroad programme recorded at the Transcription Centre can be seen as relating back to an earlier tradition that cut across the anglophone and francophone spheres of colonial influence. The radio voices of the 1960s, which went out to listeners on African and European stations, must have performed a role of dialogic connection and an appearance of simultaneity, similar to the journals of black internationalism in the years between World Wars One and Two of which Edwards writes.

Edwards argues that journals such as the Paris-based *Les Continents* and the North American *New Negro*, the paper of Garvey's globally important UNIA (see Introduction), were part of 'the transnational contours of black expression between the world wars' (2003: 3). The European metropole, particularly Paris, had a special role in the cross-flowing and buffeting of black diasporic ideas and artistic expression. Edwards draws on Raymond Williams' comment on the European metropole after World War One as a 'special sort of vibrant, cosmopolitan space for interaction that was available neither in the United States nor in the colonies'. Williams points particularly to Paris as a site of 'exceptional liberties of expression ... [W]ithin the new kind of complex and mobile society, small groups in any form of divergence or dissent could find some kind of foothold' (Williams 1989: 45, cited in Edwards 2003: 4). Edwards shows how the writings on each side of the Atlantic aimed to chart another discourse from the dominant one of international

[23] Anna Hamlin, personal communication, 23 September 2007, Melville, Johannesburg.

politics set out through the overbearing voices of empire. It was not a smooth process. There were frequent misunderstandings and moments of mutual suspicion, but the process of establishing an alternative discourse that linked the artists and thinkers of the black diaspora was established. In this period, the works of W. E. B. Du Bois, the later writings of Alain Locke, of the Harlem writers, among them Arna Bontemps, Claude McKay, Countee Cullen, and of Langston Hughes all featured.

Also important in the next decades were the libertarian transnational debates in the immediate aftermath of World War Two. Writers on American race politics in the post-war era commented repeatedly on the global dynamics the war unleashed as it brought to the forefront of black American politics a sharpened sense of an African diaspora. Penny von Eschen comments on the global spread of '"the cast of activists, journalists and editors" in papers such as the *Chicago Defender*, the *Pittsburgh Courier*, the *Crisis* and the *New York Amsterdam News*' (1997: 8). These, she tells us:

> formed a dense nexus with journalists and publishers from London to Lagos and Johannesburg, marshalling the resources of important black middle-class and entrepreneurial institutions to create an international anti-colonial discourse. From Pittsburgh to Lagos and Chicago to London to New York to Johannesburg – the nodal points of production – print journalism both provided the vehicle for the creation of this imagined diaspora and unified intellectual activists across the globe. (von Eschen 1997: 8)

Subsequent developments included the African-American participation in the 1945 Manchester Pan-African Congress and, a decade later, in the Congress of Black Writers and Artists in Paris, which took place in 1956. Baldwin's essay on the congress, which features as 'Princes and powers' in his collection *Nobody Knows My Name* (Baldwin 1960: 24–55), is quoted from by Nkosi in one of the Transcription Centre debates, to which I refer above. In the Nkosi discussion, the nub is: what, if anything, binds all black people together? On that occasion, as in the debates in Paris in Les Halles des Hommes in 1956, there was no agreement. Nkosi, like Baldwin, argued for history as the bond. It was the Ghanaian Willie Abraham who took a different line and stressed the unity of culture, taking his pan-Africanist stance from the vision of Kwame Nkrumah (Abraham 1962: 2–4).

The recorded discussions of the early 1960s produced a flow of ideas, expressive forms, and, at times, odd juxtapositions of ideologies similar to those of the journals of the 1920s and 1930s – and the later flow of print articles about which von Eschen writes. In none of these periods was there a flow of translucent diasporic understanding, hence the necessity of what Edwards, on the inter-war period, terms 'translation', as black intellectuals – both francophone and anglophone – attempted to arrive at 'a grammar of blackness' (2003: 5). There was always a need to understand the ways in which the cultures of black internationalism were 'translated, disseminated, reformulated and debated in transnational contexts marked by difference' (ibid.: 7). This need for 'translation' was equally true of what I have called the 'Nkosi period' of speech and recorded debate and the presentation of the play of cultures and power in the Transcription Centre's Africa Abroad programmes in 1962 and 1963.

Most of all, what the conversations and debates held in the taped programmes of the Transcription Centre show us is how the creative artists, politicians, and thinkers caught up in the nationalist movements of the continent sought something larger and more expansive at certain moments. They reached out to the diasporic links that cut through nationalism and suggested possibilities of very different transnational ties. The Transcription Centre tapes show us conflicting processes in operation as the talk flows: the framings of national literatures; the search for an 'African literature'; and the diasporic presence that pulled together the old continent and the Negro Americans of the New World. The radio voices, archived but still fresh and 'present', remind us of the continent's soundscapes over time and in the knot of quite agonistic struggles to be, to understand, and to make. In another sense, this was part of the linked mediality of three continents, capturing their interwoven histories of difference. The tapes leave a record of a moment of brief reconfiguration in which Nkosi was participant, broker, maker, and interpreter– of Africa 'on the rise'.

Part III

Drama, language, and daily life

5 Untidy boundaries, restless identities

Zulu serial drama in the 1970s

> For me, the 1970s was the best decade of all for the radio dramas.
>
> Nakanjani Sibiya[1]

Born from cultural and political duress

I argue here that the *imidlalo yomoya*, through the use they made of registers and convergences of voice, can themselves be seen as a resistant medium, one that mediated modern sensibilities and gave voice to new ways of seeing oneself within the family structure and interacting across generations. In particular, two of the radio dramas I discuss here show a modern view of women's roles as modern subjects. The presentation of the intimate self through the dramas also suggests a new kind of thinking about the self in the public space. This anticipated the free spaces to come in the democratic 1990s, and showed the dramas moving into different understandings of affect and interpersonal relations. The authors of the dramas also sometimes took older expressive forms and twisted them into new situations within the altered domestic space. Affect and individual creativity worked in a different psychological space, one that was not predictable but that operated in a new area of the social and interpersonal. This has resonances with the shift described by Laura Kunreuther as Nepali FM stations showed voice as the medium through which the private intimacies and the public space were configured in new ways in democratic Nepal in the 1990s (2014: 16–25). By the 1970s, the genre of *amastori* had established a large and loyal audience both in Natal and on the Rand, the populous industrial and gold-mining centre of South Africa, with Johannesburg at its heart.

Dramas on the airwaves, one of the new forms of culture produced regularly for Zulu-language listeners in South Africa from the late 1950s onwards, grew out of cultural and political duress, and thrived. This was because they soon acquired their own unique voice and style of addressivity. They drew on the appetite of the listeners for a kind of oral textuality

[1] Interview with Nakanjani Sibiya, 2011.

that mirrored a world they knew and felt familiar with, and for a voice that was not that of bureaucratic state authority, but a register which held the playfulness, intimacies, modes of address, and social self-knowledge with which they could identify. Kunreuther discusses the ways in which the voice 'exemplifies processes in which we come to recognise ourselves through something or someone outside ourselves' (2014: 254). In a way, this was what was happening through the palimpsest of voices mediated by radio drama. It was a shifting, always engaging recognition of 'the self' in an emergent public space that sometimes operated almost in parallel to the apparent public sphere of passes, segregated urban spaces, and a host of alienated places in which the black subject had to operate.

By the 1970s, the plays were being aired daily in short segments spliced together with bridging music and running for an average of 30 episodes. The sites of the plays moved between the urban and the rural and the in-between, producing a socioscape (Anderson 1991) with a thick sense of variety and a palimpsest of places. Their storylines were often finely detailed and intricately involved. Most dealt with the family unit. This alternative socioscape of the airwaves meant that disenfranchised black radio listeners were often constructing new social identities for themselves and new subjectivities as they listened to football commentaries and music programmes as well as the radio dramas, known widely as *imidlalo yomoya* (literally, plays of the air) or *amastori* (stories). I argue here that the element of control and social engineering of which the SABC and Radio Bantu were capable had its limits. Beyond the ruthless manipulation of the music industry and of music for black listeners – which without doubt existed (Coplan 1979; Hamm 1995), although it did not manage to silence black popular music in its entirety, as programmes by Guybon Mpanza on jazz and others showed – there was a space of discursive freedom, a place for reinvention in which new identities could be explored and old truths re-evaluated. Radio drama in isiZulu became an important marker of such a space.

My focus in this chapter is on three Zulu radio plays from the 1970s, a decade when African-language radio in South Africa was ostensibly firmly in the hands of the SABC, dominated then by ideologues of apartheid (Tomaselli, Tomaselli, and Muller 1989; Nixon 1994). And yet, according to the writer Nakanjani Sibiya, known for his novels and short stories in isiZulu and for his post-apartheid radio dramas on Ukhozi FM, the decade of the 1970s moving into the 1980s was:

> Very rich. When [the drama] was on, everything would stop, even in offices, all would stop ... teachers too. [And] you could stop anyone and ask about '*istori*' and they could fill you in – someone who worked in the Post Office, or a clerk, or prison officers or people in the taxi ranks ...[2]

[2] Ibid.

Clearly, something important was working to produce a resistant and aware popular culture through the medium of Zulu radio drama, which to a certain extent operated outside the borders of those who thought they were in control. The dramas within the wider station itself gave voice to certain preoccupations of a shifting and dynamic community; the broad swathe of listeners and the station itself each defined a presence through the dramas and those who voiced them. I explore here how the serial plays or *amastori* show the different faces of a radio genre that by now was loved and established. What emerges reveals the ways in which the deep tensions and resistant creativities of the era manifested themselves in the multiple voices of the dramas. I argue that the writers, actors, and producers of radio plays were working with different emphases and playing with different strands of the social and national imaginaries to which radio and radio drama gave access. Radio for African listeners, in the form of the station known from 1973 as Radio Zulu, was envisaged as a means of controlling a restive, modernising black populace. Yet, as a technology of modernity and through the figure of the voice, it helped create modern subjects.

The South African state had itself embarked on a modernist and modernising project through its haphazardly thought-through apartheid policies (Posel 1991; Hyslop 1993). Jonathan Hyslop (1993: 397) shows how the National Party's policy of Bantu Education (begun in the early 1950s) was a massive operation in social control, particularly of urban youth. More than that, through this 'ruthless' project of Bantu Education, 'the Nationalist government was able to create a "stable" apartheid system from the early 1960s until the end of the 1970s' (ibid.: 406). This was a project of policing biopolitics (Foucault 1978) and enforced 'modernisation' among its African population within the often reactive and self-contradictory race policies of the era. The modernity that radio drama helped create was of a different kind. Radio drama in isiZulu proved a perversely resistant medium and genre. The resilience and restlessness of the language itself, the addressivity of its genres, and the creativity of producers and consumers alike made it impossible to control or even monitor with more than minimal success, as writers on Northern Sotho Radio and on Xhosa radio drama – both of which, with other languages such as ChiVenda, followed from the model established by Zulu radio drama – have also pointed out (Gqibitole 2002; Lekgoathi 2009; 2011).

In drama particularly, the intimacy of the medium allowed a concentration on emotion and interior life that was crucial in the creation of modern subjectivities. In addition, the dramas, in their new configuration of emotions, could cut across entrenched cultural expectations and present differently engendered 'emotion discourses' (Lutz and Abu-Lughod 1990; Abu-Lughod 2000). The focus in many dramas on the family and on ordinary life was another part of the construction of the modern. In their frequent focus on the family and the self, the dramas produced a discourse of the modern related to social life. This in turn brought about its own alternative versions of power, community, and selfhood, which quietly – or perhaps noisily – coexisted with the wider public production of apartheid-driven discourse.

The writers of the scripts, who were often also producers and actors, as well as broadcasters were able to set out what Charles Taylor in his *Sources of the Self: the making of modern identity* has called a 'sense of the importance of the everyday in human life' (1989: 14). This affirmation of the production and reproduction of everyday life, Taylor regards as 'one of the most powerful ideas in modern civilization' (ibid.: 14, 23).

Space, voice, and radio

Part of the intimate interiority of radio was its ability to enable the inner eye to move over both the soundscapes and topos of the urban and the rural at a time of legal spatial restrictions for black listeners. Like the medium of print, radio could provide a means of travel. For radio, it was primarily the voice working across registers of tone, social and gender identity, regional accent, timbre, and even profession that allowed multiple journeys and passages, and enabled radio to mediate an entry into new modes of being and of becoming modern. It allowed the inner eye and ear to engage with 'tradition' and the modern, which meant that the contestations over culture and national identities present in the 1970s in South Africa played themselves out through radio in opaque ways.

Elements of 'tradition', in the form of the varied genres of the cultural archives, could be present within a drama and yet be accessed differently as they nested within a longer narrative form where they became not discrete moments of experience but parts of a longer dramatic conversation. Within such a frame, for instance in the first play I discuss, *Ubongilinda Mzikayifani* (You Must Wait for Me, Mzikayifani), listeners could chew over segments of praise poetry (*izibongo*) and assess them as part of the expressive whole offered by the characters within the unfolding drama. The classic expressive genres such as praise poetry thus became part of the interpersonal and family tussles in which listeners were involved over many weeks. The older genres related back to established kinds of performance tied to sociality, memory, and history. Yet they also became an element in the new medium, held in a web of associations that linked character, narrative development, resolution, and thoughts on a moral order relating to 'the production and reproduction of everyday life', as Taylor puts it (1989: 23).

Voice, in the material form of the radio, also created socioscapes, or what Arjun Appadurai (1991) has termed 'ethnoscapes'. These, through their access to new images of social life, spread by the media and 'complex transnational flows' (ibid.: 209), undercut the restrictions on black movement that were a cornerstone of segregationist laws. The slender resources allocated to the African-language service of Radio Bantu intensified the multi-skilling common to radio in places such as the BBC (Hendy 2000), sometimes with unexpected results. The limitations on black radio in terms of budget and staffing meant that voice created a complex set of memories, loyalties, and imaginings, as broadcasters constantly doubled up, playing different roles and taking listeners to different social sites.

Koos Hadebe, who joined the Johannesburg SABC in its downtown Commissioner Street offices in 1974, was one of four full-time staff for the Zulu service. Together with Thetha Masombuka, he was football commentator, covered music programmes, and acted in serial dramas. Voice was a crucial element of the circulatory power of the new medium – and of drama within it. Not only was the station and its small team – strung between Johannesburg and Durban – the beneficiary of the early struggles of Masinga, Sishi, and others to have a voice for isiZulu radio, they were also able to give voice to the concerns of listeners. Much of this happened through the prism of the dramas and the struggle to resolve problems of affect, loyalty, and so on that had value in the day-to-day lives of listeners. On a micro-level it created strong and complex loyalties as listeners began to identify with a particular personality through his or her voice. The layers of archives that were built up from the multiple personae of a particular person as a character in a drama, as a news or sports presenter, even as a producer, were part of these loyalties and became a way of shaping radio culture in ways not expected by the initiators of Radio Bantu. They became a means of rooting lives in a modernity constructed against the grain of the ideologues. Listeners could travel with the same voice. It could take them to known physical locations such as the football stadiums around the Rand in Daveyton, Benoni, Witbank, Attridgeville, and Mamelodi, where teams such as Witbank Aces, Moroka Swallows, Orlando Pirates, and Kaiser Chiefs battled for supremacy.[3] It could also transport them to sites of the imagination through dramas that listeners inhabited and took over as their own. Thus, voice, as radio soundings, made it possible for listeners to build up a web of associations across programmes and over time and place. It produced different discourses of the social, of interpersonal relations, of the family and home, even of the nation, and thus it produced its own social power, and new subjects.

For some listeners, time and place knotted together as key radio moments – these often related to football, radio drama, or both – and

[3] Interview with Koos Hadebe, 2000.

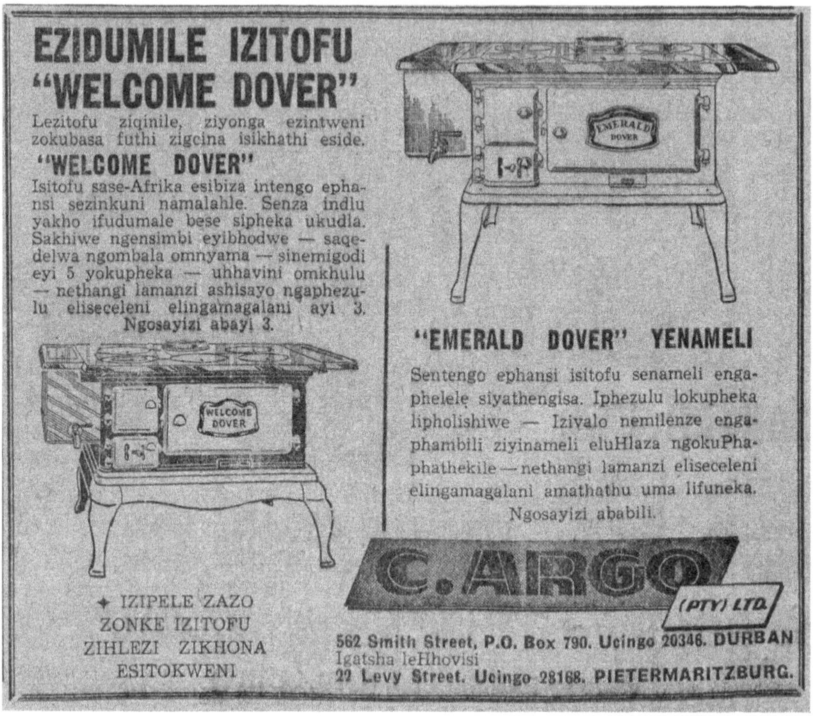

Figure 5.1 The famous 'Welcome Dover / *Ezidumile izitofu*' *Ilanga* newspaper advertisement, January 1960.

these events shaped the daily or weekly rhythm of their lives. In 1970s South Africa, radio and the intimacy of urban or peri-urban domestic life were often linked with a certain key material item – the famous Welcome Dover stove, with its long chimney going outside, its four top plates, two side doors, one for the wood and coal to heat it and the other for cooking, and its armoury of cleaning implements in a shelf beneath the cooking/baking compartment.

Take, for instance, moments from the boyhood of Tinyiko Maluleke, now a senior academic whose voice is often heard as a political commentator on SABC radio stations and on television. Let us move to Soweto in the early 1970s – to Meadowlands. It is 2 p.m. on a Saturday afternoon, in winter; two small boys are hard at work cleaning out the stove in the main family room in the standard matchbox house beloved of apartheid township planners. Why are they there and not playing outside? It is so that their task of cleaning and preparing the Welcome Dover and then lighting it and warming the house in the cold Highveld winter can be over by 3 p.m. Why 3 p.m.? Because that is when the football commentary starts and the voice of the famed Thetha Masombuka – on Radio Zulu – will

Figure 5.2 Across the generations: 'A man who owns a radio enjoys more and learns more', *Ilanga* newspaper advertisement, January 1960.

transport young and old to the grounds of contest, where teams battle it out for primacy, listened to by their fans all over the country.[4] One did not need a pass, or money for travel through apartheid's mapped cities, to move in the mind to that pitch, spurred on by the fevered eloquence and soaring tones of Thetha Masombuka as he created verbal pictures of skill, daring energy, spectacular tries, near misses, penalties, fouls, off-sides – and in the process built up images of future culture heroes and lauded the prowess of beloved teams such as the Swallows, Chiefs, or Sundowners.[5] These were images of what could easily become a proto-national pitch, a binding together of listeners into an imagined community (Alegi 2010a; 2010b). Masombuka's football commentary may not have created a nationalist discourse but it made a free space, cherished at a time of heavy censorship and restrictions on the movement of black people.[6] Among the commentators themselves there was a sense of a public constituency and of 'serving the nation', partly because of the range of listeners drawn from the different language groups, and also through a sense of 'the sheer number of people who depended on you'.[7]

Running alongside the multi-lingual, multi-ethnic shadow nation that football commentary on the Rand called into being was the intense focus on the domestic that came from the serial dramas. The listeners for these were both urban and rural. They communicated with producers by mail at the end of each drama and their views, even when seemingly perverse, were treasured by the producers, dramatists, and actors as intimate ties with an unseen yet present community of listeners.[8] It was within the site of the dramas, I argue, that a new kind of modernity was shaped.

Voice, space, and the domestic

The radio serial *Ubongilinda Mzikayifani* played for 20 episodes on Radio Zulu in 1974. It was produced in the cramped Radio Zulu studios at 1 Commissioner Street, then the hub of downtown Johannesburg, with a number of well-known names among the cast. Produced and written by Bhekisisa Kunene, who was also a frequent newscaster,[9] it included Wiseman Mkhize as the unsuccessful suitor, Sokalezwe. Minor roles were played by Alexius Buthelezi, by then a household name in the black world of choral music and radio, and three other well-known actors: Johnny Magagula, Sagius Thwala, and Sophie Mgcina. Elijah (Thetha)

[4] Interview with Tinyiko Maluleke, 2000.
[5] I am grateful to Tinyiko Maluleke for sharing these reminiscences of his boyhood days with me. Interview with Tinyiko Maluleke, 2000.
[6] On sport and nationalism in South Africa, see 'Apartheid on the run: the sports boycott' in Nixon (1994). See also Couzens (1983: 193–214).
[7] Interview with Koos Hadebe, 2000. [8] Interview with Winnie Mahlangu, 2006.
[9] Wiseman Masango, personal communication, 7 July 2000.

Masombuka (famous also as a football announcer, as noted above) played the part of Mzikayifani, the lyrical yet sensible wooer of the resourceful young woman, Nomgcagco, played by Idah Nkosi, whose words are the play's title: 'You must wait for me, Mzikayifani!' What was key was that a star football announcer – as voice – and the star male character in the radio drama converged, giving, I suggest, an extra dynamic, a sort of double weight, to the power of the mediatised voice.

The play used the stereotypical traditional setting of a courtship tussle. Two youths vie for the affections of an attractive young woman, with much of the action taking place at the river where girls go to draw water for the family's needs – the river and courtship being a key site of cultural memory and affect referenced by musicians in modern songs that invoked both the present and memories of past custom. The drama, though, worked with a play of emotion more widely and with more interiority and narrative length than would be expected in the classic oral genres on which it drew, or in a modern song such as Mafikozolo's great hit 'Meet Me at the River' (2005). It thus set the cultural archives upon which it partially depended within a different configuration. Lila Abu-Lughod (2000: 96, 99) has pointed out the striking differences between the emotional styles and imaginaries created by 'traditional' Egyptian and Arab forms of cultural expression and narrative and poetic traditions and those of TV melodrama. She tells us that the epic poem about Abu Zayd al-Hilali, when transposed to television, becomes a melodrama:

> About interpersonal relationships and individual longings and passions set in the domestic sphere . . . Instead of formulaic phrases about tears and their plenitude, what could be thought of as the phenomenology of emotion, TV drama tries to produce their inner beings . . . through close-ups of facial expressions and melodramatic acting.

Something similar but different happens in *Ubongilinda Mzikayifani*. The older genres are put to work in new ways and move in a different pattern of affect. This is particularly true of the women's 'traditional' genres in which confessional emotion has always had a place. The classic women's genres of Zulu poetry include the intimate, emotional ballad style of the famous poem of lost love *Nomagundane* (Miss Rats), which was recorded in the 1960s by Princess Magogo of the royal house when she sang it for David Rycroft (Rycroft 1975; 1975/6) and at times for her own family members. This style allowed for interiority and for individual emotion and dramatic development, and can be seen as a very modern genre, a forerunner of the radio dramas. Other classic women's genres include the shorter personal yet formulaic love songs of young girls (Joseph 1983; 1987), the autobiographical, confessional mode of praise poetry for children (*izangelo*), and women's praise poetry (Gunner 1979). These, beautifully, allow space for individual emotion and very condensed, often highly allusive, narrative.

However, with the exception of the remarkable *Nomagundane*, the shorter poetic styles allow little room for any narrative development or any extended setting such as that allowed in a serial drama. Yet *Nomagundane* has the seeds of a modern radio drama: a young woman who has been jilted by her lover, perhaps under pressure from his mother, describes to her sister, Nomagundane, in slowly unfolding details how this happened. She relives scraps of dialogue and sketches actions between the principal players, which gradually become significant as events unfold to their final denouement of rejection. In the ballad, there is no reassertion of a moral order, which almost invariably marks the radio dramas, yet the similarities between the two forms are striking. The dwelling on emotion and interior life, the sketching of an intense, individual subjectivity, is a central dynamic of the ballad. This is what the serial builds up, and, in fact, propels into a much higher dynamic when it makes use of oral genres. It expands the known forms, makes them elastic, and at the same time sets them in a much wider frame of interpersonal relationships, family relations, intrigues, and a new context of gender politics.

In *Ubongilinda Mzikayifani*, domestic and interpersonal relationships come under the closest scrutiny. The older genres that the drama brings into play are those of men's poetry, not women's, although the latter can be understood as silently present. The two young men who are in competition display different kinds of masculinity: one, Sokalezwe, is boastful and fluent in his self-praises (*izibongo*). He is, as his name proclaims, 'the Nation's Best Bachelor'. His rival, the winner in the love triangle, is Mzikayifani (Households Are Not All the Same), who, the serial makes quite clear, will make a far better husband and lover. The chief claim of the drama to show the move to inner life, interiority, and hence the modern subject comes through its emphasis on the trials of young love, set within the home and the family. A related modern focus is on marriage, love, constancy, and equality. Also present are the voices of a strong mother and a determined daughter. They find a way through misfortune and turn dire events to their favour by dint of the energy of speech and, in Nomgcagco's case, through an intense hold on what is right even though it means following a course of action that is not always transparent.

Thus, in the same year – 1974 – when Thetha Masombuka was holding his listeners' rapt attention with his football commentary, he took the leading role in what was to become one of radio drama's small classics; it was aired again in 2000, 6 years into the post-apartheid period and 26 years after its first broadcast. In this 20-episode play, short in comparison to most, Masombuka played the role of Mzikayifani, the lovesick young man whose sweetheart inexplicably turns him down, saying, enigmatically, 'You must wait for me, Mzikayifani!' On one level, *Ubongilinda Mzikayifani* could not be further from the hectic world of football. Its setting is deeply rural. It exploits the old and enduring cultural tropes of courtship at the river, or stream, with the young girls continually going to draw water and being waylaid by their admirers. Yet within this almost formulaic narrative there are finely drawn variations. Mzikayifani, the ardent suitor – also Thetha the announcer – shows the emotions of equality, not conquest. A blueprint for the sharing of sorrows and burdens is set out in his address to her. When Nomgcagco (in Episode 2) shares with him her anxiety over her father's illness, he answers:

> No Nomgcagco! Indeed, we would face anything terrible together, whatever happens – we would bear the sorrow of [his] illness together. It's always like that. The sorrows of the world bring people closer. You would never expect, surely, to be made to carry the grief of your house alone? We are here, with our love. Nomgcagco, surely you see that your sorrow is my sorrow too? That's why I say Nomgcagco, give me your word that we will indeed be one. May nothing of this world part us![10]

Yet Nomgcagco, as a woman of action, contained emotion, and few words, has little time for Mzikayifani's eloquence. His finely turned phrases and her forceful brevity bump against each other. 'My! Are you a child! Haven't you taken in what I've just said to you?' she replies.[11] As the complicated plot unwinds, we see Nomgcagco forced to pretend to give her love and promise of marriage to Sokalezwe (Nation's Best Bachelor). It is he who has stolen the medicine bag of her father's herbalist doctor (*inyanga*), making it impossible for him to cure her father. If Nomgcagco agrees to marry Sokalezwe, he promises he will return the bag to its rightful owner, her father's doctor. Meanwhile, Mzikayifani's language increases in intensity and lyricism as he presses his case, always at the river. In Episode 7 he draws on the beauty of the natural world, and, with a lyricism reminiscent of Elizabethan sonnets, he paints a picture of the pastoral context of their love. This time, his words work strongly on Nomgcagco's emotions:

> MZIKAYIFANI: Come ... let's go and sit under a willow and listen to the rustling of the leaves, singing a song for us, Nomgcagco, that we'll never part. We'll hear the birds above; we'll see the white egrets flapping their wings in their flight to tell us that the birds of the air approve of our everlasting love. Then, Nomgcagco, we'll feast our eyes on the little 'water boatmen' skimming and dipping ecstatically in the water. And the water itself will be babbling on,

[10] Bhekisisa Kunene, *Ubongilinda Mzikayifani* (1974). Transcribed from the tapes of the radio drama by Wiseman Masango and translated by Liz Gunner.

[11] Ibid.

singing a song of mourning as it leaves the pool and tumbles down a deep hollow. Come closer Nomgcagco, Nomgcagco!

NOMGCAGCO: I do feel your words Mzikayifani.[12]

As the two strands of the drama move forward, the sick father seesaws between near death and a return to health, and his doctor frets and fumes at the loss of his potent medicine bag. Mzikayifani pursues his courtship of Nomgcagco – always at the same high level of eloquence and intensity; she listens, her emotions held in check. She schemes how to retrieve the precious medicine bag from the devious and persistent Sokalezwe.

By Episode 17, Nomgcagco, to the dismay of her mother and the girls of her age group, has agreed to marry Sokalezwe, accepting him by saying, ambiguously: 'I have felt your sting, child of a great man. Words fail me.'[13] The drama moves swiftly to its climax and denouement in the final two episodes. Nomgcagco's father's health improves and his doctor is able to use his powers again. He sends a terrible storm to destroy the home of whoever stole his medicine pouch. News comes of the demise of Sokalezwe's father as his home has been struck by lightning and burned to ashes. So it is Sokalezwe – boastful, unreliable, and unscrupulous – who is punished for his selfishness and greed. Nomgcagco is rewarded for her courage, loyalty to her father, and general virtue. Mzikayifani is also rewarded for his steadfastness in his pursuit of Nomgcagco, and, it would seem, for his amazing way with words, his command of language – and perhaps for the equality of intimacy that his words seem to promise.

This drama shows the scope of the rich and lyrical language of courtship and exploits to the full an ornate vocabulary steeped in images of bountiful, harmonious nature. It constructs an enclosed world yet one with its own interior dissonances and moral threats. It holds in its setting no whisper of a modernising, restless social order. Yet, in dwelling on the emotions and the interiorities of its two chief characters – the youth Mzikayifani and the resilient, compassionate Nomgcagco – it is intensely modern. Thetha Masombuka's voice, which shifted between the passionate eloquence of the character Mzikayifani in an unnamed rural place, and his own rich depictions of action, style, physique, and player biographies in the stadiums of the Rand, was itself a point of mediation for the multiple worlds in which people lived and negotiated their identities.

There is another way in which *Ubongilinda Mzikayifani* may have held a deep fascination for its listeners, and depicted a version of black modernity. Although intensely modern in its emotional framing, while drawing on the classic genres, it made use of indigenous cultural archives in its depictions of sickness and health. The herbalist doctor's power to heal Nomgcagco's father and to play a part in re-establishing a moral

[12] *Ubongilinda*, transcript 8. [13] *Ubongilinda*, transcript 17.

order that had been displaced showed a different biopolitics being brought into play. The healing of Nomgcagco's father and the retribution for the theft of his doctor's medical knowledge were not brought about through Western medicine. It was knowledge and power that drew unambiguously on 'tradition'. This can be seen as a kind of cultural translation, as the modern – and a particular kind of modernity – is given shape in the drama. As Timothy Mitchell remarks in relation to an essay by Dipesh Chakrabarty on shifts in Bengali culture in response to a Western, colonising modernity: 'The modern subject emerges not through the simple adoption of European categories of thought but through a continuous process of cultural translation' (Mitchell 2000: xvii).

This particular process of translating the modern to the huge body of drama listeners – listed for us by Nakanjani Sibiya as clerks, teachers, post office workers, taxi drivers, and so on – may be one reason why *Ubongilinda Mzikayifani* played again on the airwaves in 2000, as the post-apartheid era contemplated a new set of demands and imaginaries. The times called for an inclusivity of imagination, a 'continuous process of cultural translation' like that presented in *Ubongilinda Mzikayifani*. So, on one level, the drama captured in timeless 'tradition' was in fact a space, through its voices – authorial and acted – for the staging of a modernity that spoke directly to the dilemmas of love and interpersonal relationships, as well as to the values of older moral orders. These later listeners constantly engaged with something new, even as they negotiated the changes. In a way, the voices opened up these spaces both directly and yet also at one remove. Most importantly, the female characters – in particular the mother and her daughter, Nomgcagco – moved boldly and surely among moral difficulties, uncertain emotions, and demanding men. At the drama's heart, therefore, was the family and the intensity of the everyday in the modern world.

Women without a home: *Abangane Ababi* (Bad Friends)

Although working within certain prescribed frames, one of which was an almost invariably linear narrative line, many of the scripts were relentless in their exploration and, in a perverse way, celebration of the life and language of the times. In a way, the *amastori* could sound out possibilities and dangers, again through timbre, modern idiom, dodgy characters, and women skirting the edges of convention. *Abangane Ababi* did all these things. It ran from early May 1979, and on one level is a modern fable of a young girl who narrowly escapes corruption and disgrace. Only at the very end of its 36 episodes is a degree of moral and social equilibrium restored. The play, by Abigail Zondi, and again produced in the Johannesburg studios at 1 Commissioner Street, is a discussion of the options for young women, in a hybrid and modernising world, and the risks of these options.

The segments, with bridging music chosen by the producer and often mirroring the emotional states of characters, knit together the border territories of place and morality. The drama is set in the south of Natal, near the tiny hamlet of uMkomazi and the coastal town of uMzinto. An anonymous and unidyllic rural space, this is the place from which the young woman, Nomathemba, tries to move away. There is also a black apartheid township setting and a shebeen run by a free woman of uncertain reputation called Aunty Sarah. Importantly, there is a group of 'mothers', eloquent, assertive, and constantly anxious about the behaviour of Nomathemba. Their voices are heard within the circle of women, so they do not need to speak softly or quietly, as they might in mixed company. Finally, there is Nomathemba's mild, good-natured, and dull fiancé, whom the author keeps mostly offstage until the plot needs him, when vice fails and virtue is restored. Shadowing the naîve Nomathemba is the worldly wise Danisile (whose name means 'Disappointed'), who has recently returned home after a failed marriage. When she urges Nomathemba to go to the 'bioscope' (films) with her at nearby uMkomazi, saying that, after marriage, she can be sure that her husband will leave her at home and take other women to the pictures, Nomathemba is intrigued and interested. The drama thus explores the difficulties of marriage in either a 'traditional' or a more modern setting.

Abigail Zondi was one of only a few women writers for radio at the time. The politics of gender works itself out in the play's language and in the range of its female voices. It is full of racy, domestic humour that echoes the feisty banter of women's praise poetry and the energy of daily domestic dialogue. 'Am I the only one to be defeated by marriage? I'm going! Don't chase me out like a dog that's been eating eggs!' Danisile says to the 'mothers' defiantly, as they hustle her from the house when she is visiting Nomathemba. Danisile, although treated sympathetically, is not held up as a role model. Unlike Danisile, Nomathemba has a strong and close relationship with her mother, but she frets that her mother continues to treat her as a child. She explores the border spaces beyond the rural home and begins to work for the shebeen queen, Aunty Sarah, finally living with a shady character, a man of the streets, with little money and little work, called Mjika Jo – a name still favoured by taxi owners and painted, slogan-like, on their vehicles.

Affect in *Abangane Ababi* is vested not so much in a love relationship as in the encounters between generations of women and between a young woman and the brawling border world of bars – the modernising and in-between spaces of the 1970s. It becomes a discourse of the social (Lutz and Abu-Lughod 1990). The drama gives voice to the range of registers in modern isiZulu, and to the hybrid nature of the modern

African culture in which it is embedded. On the bus, and at another point in a taxi, an Indian character speaks Fanagalo (a basic, mixed-up isiZulu made popular in the mines); Mjika Jo speaks a Zulu city argot, throwing in still popular phrases such as '*Ek se!*' [Afrikaans for 'I say!']. Finally, Mjika Jo and Nomathemba fight over their gender roles as they sell beer in their shack:

> 'You get up and pour for them!' says Nomathemba. 'No!' replies the conservative but not 'traditional' Mjika Jo: 'I'm a man!'

At the climax to the drama, forgiveness and the restoration of order take place in hospital, 'in Ward A', where the badly beaten up Nomathemba is recovering, with her anxious fiancé, Magwaza, and the 'mothers' at her bedside. The ending offers one option for a moral order, yet it is set unforgettably within a tough, modernising, subaltern world that would be familiar to many listeners as a space that needs to be included in such a moral order.

Abangane Ababi thus takes its listeners into the unmapped spaces beyond set cultural norms, again using a dense register of modern Zulu and a mix of characters. It shows young women navigating the tricky spaces for choice in interpersonal relationships and freedom. The family becomes the point of both departure and uncertain return as Nomathemba moves into an emotional and moral terrain about which the cultural archives have little to tell her. Interestingly, however, the 'mothers' – in the form of her own loving mother and her friends – play a prominent part in binding her to them through love, as the play ends with Nomathemba, battered by experience, in her hospital bed in 'Ward A'. Here, then, is a mid-1970s serial drama where the company of women and the voices of women dominate, and one that allows new women's subjectivities to emerge.

Control at all costs: *uBhekifa* (Look After Your Inheritance)

Not all the Zulu radio dramas of the 1970s sidestepped the preferences of the apartheid mandarins at the SABC. Listeners and writers/producers also had to contend with apartheid ideologues and uneasiness about incursions from the north and the broadcasts of Radio Freedom, which was operating from Dar es Salaam at the time (Davis 2011). On one level, *uBhekifa* is nothing more than dramatised propaganda. It mines the cultural archives and sets them in a narrative that conveys a message of fear of outsiders. It reproduces a static picture of 'native culture' at risk from dangerous outsiders. It constructs a narrow ethnicity entirely in keeping with the apartheid blueprint that lay behind the establishment of the African-language radio stations of Radio Bantu. Yet it might also have been so polysemic that its driving message of fear of the outside world was submerged within a wider fascination with place as a metonymic ideal of nation. This serial drama shows that the crude propaganda so blatant in the Zulu-language radio news items (Scott 1999) could also be presented as drama through metaphor and narrative, and with recourse to the aural resources of the new genre. Even here, however, language and the palimpsest of voice proved a resistant medium that could prompt multiple responses rather than a single predictable one.

UBhekifa takes its name from one of the central characters, a policeman from the quiet city of Pietermaritzburg, north of the city of Durban. It was on the air for 40 episodes, over approximately 8 weeks, from mid-March 1970. This drama must have been intended to scare black youth in particular away from embracing the radical oppositional positions that were growing in strength throughout the country at the time. It would also be a warning to others of the 'dangers' lurking to the north of South Africa's borders. There was much for the South African state to be anxious about, particularly with regard to youth:

> Coming of age politically in the 1960s, SASO's [South African Students Organisation] founders had watched most of Africa pass rapidly through the transition to independence, including the neighbouring territories of Botswana, Lesotho and Swaziland. By 1970, the Portuguese were fighting unwinnable colonial wars in Angola and Mozambique, and Ian Smith's rebel government in Rhodesia was under pressure from world sanctions ... In South Africa the literature of African anticolonialism was supplemented by collections of the post-independence speeches of Julius Nyerere of Tanzania on the theme of *ujamaa* and African socialism ... Also influential were the writings of the Martinique-born psychiatrist, Frantz Fanon ... Fanon ridiculed colonialism's self-serving definitions of right and wrong ... Feelings of anger and hatred in the oppressed, he said, were not symptoms of moral weakness but healthy responses to the indignity of being treated as subhuman. (Karis and Gerhart 1997: 104)

In 1970, a large (exiled) ANC delegation visited the Soviet Union and held lengthy talks on 'irregular warfare'; it was also the year before 'Operation J' attempted – unsuccessfully – to land 50 armed guerrillas on the Transkeian coast (Karis and Gerhart 1997: 51). The fears of a black uprising, of infiltrators corrupting a supposedly docile and innocent indigenous populace, dominate the drama *uBhekifa*.

In addition to the subaltern and enriching modernising of the Zulu language itself that marked the 1970s, it was also a period when new Zulu words denigrating radicals began to be used frequently in SABC radio news broadcasts in the language (Nene cited in Scott 1999). One such term that features frequently in *uBhekifa* is *amaphekulazikhuni* (literally, what blows over the burning faggots), meaning Bolshevists or militant communists. Throughout the 1970s, the onslaught against oppositional forces waged by the apartheid state through the medium of radio – and in particular through news broadcasts and news commentaries – was ferocious. The African-language news services were particularly heavily policed (Gqibitole 2002; Lekgoathi 2011). In an interview with the journalist Sally Scott in the Durban-based *Daily News* on 23 August 1999, the veteran newscaster Thokozani Nene spoke of the difficulties of those years of censorship and control. He would, he said, read the carefully prepared and heavily censored news, always prefacing it with the disclaimer that he was reading not what he believed but what had been written. This became standard practice among the Zulu newscasters at the time. Nene also tried another route: he would end his readings with short snatches from the praise poem (*izibongo*) of one or other of the Zulu kings – Shaka, Cetshwayo, Mpande, Dingane. He would vary his choice of royal personage, and he would address the praise poems to his listeners: 'You, offspring of Zulu son of Malandela . . .' In the interview with Scott, he recalls that it was not long before the white authorities announced that they felt Nene's unusual way of rounding off the news was a form of incitement.

UBhekifa, like the two dramas discussed above, was produced in the SABC Commissioner Street studios in Johannesburg. It was scripted by P. V. Dlamini, produced by the illustrious and diligent Alexius Buthelezi,

Figure 5.3 Thokozani Mandlenkosi Nene, the renowned Radio Zulu and Ukhozi FM broadcaster and actor in radio dramas.
Source: *Ilanga* Photographic Library.

and the narrator was the football broadcaster Soks Kubheka, who also played the role of the policeman, uBhekifa. Other radio personalities, including the football commentator Thetha Masombuka, have minor parts. The setting in *uBhekifa* is neither idyllic and alive nor filled with the turbulent energy of the borderlands. The narrator moves listeners to the north of present-day KwaZulu-Natal, to the uBombo mountains near the border with Swaziland and Mozambique.

The outside world, in the form of either Johannesburg, with its criminals, sleazy shebeens, and communists, or the dangerous 'north', contrasts with a carefully evoked, overdetermined rural space. The opening lines, as a man returning home after years in Johannesburg gazes around him, paint a scene of extreme natural beauty and tranquillity: the trees are etched as if on a background of gold. It is, moreover, a peopled landscape. The man sees cattle and herd boys in the distance; he sees, as part of the natural order of things, the homestead of the chief of the district. His eyes follow the course of the stream called Inguya that passes below the homestead; he makes a mental note of where it flows into the larger stream of Zanezigodo and follows it in his mind's eye through the mountain gorges until it reaches the dark forest of Dukithole (Lost Calf). He gazes at the waterfall at its edge, until evening falls and he realises that he must continue his journey home. The overwhelming sense is of the artificiality of the scene, its constructedness. It is a painted, static scene, as if seen artificially from a great distance.

The careful marking of the landscape, its great natural beauty, and its over-precise mapping of the social order is set against a symbolic tale of ogres and hyenas that emerge at night from the forest of Dukithole to eat innocent children. This is a tale to which the returning traveller listens as he sits waiting for the chief. He listens, a stranger on the edge of events, as the children cluster around the grandmother narrator who is a consummate storyteller. This again is a set scene of what 'native life' is about. Clearly, the ogres and hyenas that hunt in packs and eat children represent the intruders, the *amaphekulazikhuni*, who, we are soon told, have asked the chief for refuge from the police and are living in the dark forest of Dukithole.

The plot, running over 40 episodes, twists and turns but maintains suspense and intrigue. The strangers, who are ruthless once it is clear that the chief will not assist them, attempt to set his people against him and threaten to destroy him and his community. In another piece of ethnic programming through narrative, the strangers are not mother-tongue Zulu speakers. When the young boy Masilo, who has listened fearfully to his grandmother's story of the ogres and hyenas in the forest, encounters three strange men as he struggles to free a trapped calf from a hole, they speak to him as brutal outsiders. The leader talks to him in isiXhosa. Simple Xhosa phrases, roughly spoken, are thrown in as markers of difference: '*Heyi kwedini, wenzantoni apho*?? *Thetha*! *Thetha masinyane*!! [Hey young boy, what are you doing there?? Speak! Speak quickly!!]' The other two men with the Xhosa speaker are even more alien and come from 'a country to the north'. The outsiders, like the ogres of the folk tales, live in the forest and the caves within it and are again marked as the 'other'. As the serial unfolds, the strangers are progressively 'othered'. They are from Zambia; they are called, by Episode 10, '*amaphekulazikhuni zamakhomanisi*' (terrorist lackeys of the communists), and the stage is set for a battle between the forces of stability – those of the *inkosi* (chief) Bhekumuzi Zondi – and the *amaphekulazikhuni*, the terrorists. There is a great deal of suspense; there is a turncoat villain and a beautiful girl whose loyalty is in question for a long time. Has she defected to the *amaphekulazikhuni* or not? As one would expect, the terrorists are defeated, the girl is discovered to be loyal, and order is restored. The model of a dominant Afrikaner nationalism, holding the smaller ethnic consciousness in check and defeating the invading strangers, is reasserted, and the ideology of a broader unity of isiZulu and isiXhosa, which becomes a larger African nationalist identity, is crushed. There is very little family interaction; instead, there is a bleak emotional world of fear with lightly sketched instrumental relationships and no intimate domestic scenes of the everyday, other than the set piece with the grandmother telling her story.

And yet, even in this most ideological of dramas, there is a certain dialogism. This comes through language, rather than any particular timbre or register of voice. The scene of natural beauty so carefully etched at the drama's opening has affinities with fictional descriptions of the land drawn to evoke feelings of belonging and patriotism. Edouard Glissant's seminal novel *The Ripening* (1985), to take just one example, uses views of the land to evoke a deep sense of Martinican and Caribbean identity and the need for revolutionary change. The early descriptions, although fixed and static on one level, have the capacity to represent a freed country rather than one under threat from vicious outsiders. Here, too, in this seemingly most controlled of dramas, there is the interpretive capacity to move between the narrowly ethnic and a broader trans-ethnic identity. There seem to be no archival records in the SABC of the millions of postcards and letters (and phone calls) that poured in to Radio Zulu from avid listeners (Switzer 1985) as the dramas unfolded.[14] Thus, the written listener responses to *uBhekifa* are unknown. The episodes may have elicited a mixture of responses – suspense, fear, delight in the play of natural imagery, nostalgia, scorn. The memory of the drama held by one urban and 'conscientised' youth of the era, the

[14] Interviews with Eric Ngcobo, 2011, and V. V. O. Mkhize, 2009.

poet and novelist Mandla Langa, was of the manifest hostility it showed to an imagined world that could exist beyond ethnic divisions and an unchanging, set-piece 'Africa'.[15]

Present incursions

Ukhozi FM, as the former Radio Zulu is now called, is currently the most successful African-language public radio station, together with the Sesotho station Lesedi FM (Coplan 2011). Could this strong presence have been imagined by broadcasters and newscasters such as Thokozani Nene as they struggled under the restrictions of doctored news items and misinformation ordained by their superiors? What propelled them was the clear belief that in the end the energy of the spoken language, imagining and 'speaking' the utterances of people in everyday and extraordinary communication, would win through. It was the forum of voice established by radio that created not only a resistant community but also an alternative public space, *inkundla,* where voice could be both made and given form. Certainly, in the hands of their determined and creative writers, with their loyal, loving, and discerning listeners, the dramas built something unimagined by the controllers. Language, a palimpsest of voice and a 'knowledgeable community', set out its own rules. The stories of the everyday grew daily, weekly. They provided a means of constructing a version of the modern that sustained and provoked listeners and helped them navigate the journey to the present.

[15] Interview with Mandla Langa, 2001.

6 Radio drama in the time of violence

Yiz' Uvalo (In Spite of Fear), December 1986 to May 1987

It's about warlords. Khaba Mkhize [1]

It was a period of abductions, abuse, ambush, armaments, arson, assault, attacks, attempted murder, axing, battles, beheadings ... weapons, wounding and wrongful arrest. (Aitchison 2015: 65)

A new aural aesthetic of violence

This chapter has as its focus the life of a particularly long-playing, and undoubtedly gruesome, radio drama of the late 1980s. It captured the violent spirit of the time in a way that was uncompromising. It was full of terrible pain. It was also, like so many of the dramas, about the family, and 'the everyday'. The defining soundtrack, the '*ibika*' that played at the start of every episode, was from Tina Turner's hit of the 1980s 'We Don't Need Another Hero'. It showed the easy movement and criss-crossing of black music and the popular culture of the radio dramas, the *imidlao lomoya*. It showed, too, the skill of the producer, in touch with popular taste and the 'feel' of the serial drama itself. The writer was Morris Vusumuzi Bhengu; he was also a novelist of note, but had more radio dramas to his name than novels (Sibiya 2001b). More than his name, it was the characters, the tragedy, and the dark forces of the everyday that the drama released, and perhaps mirrored, that people found irresistible. In its eerie soundtrack it seemed to create a new aural aesthetic of violence; and it stuck, succubus-like, to some of those who acted in it, threatening to take over their public personas. The drama ran over two parts and was known largely by the name of the first part, which was the start of a proverb, *Yiz' Uvalo* (In Spite of Fear), and covered 40 episodes. The second, completing section of the proverb, and the title of the second part of the drama, was *Umanqob' Isibindi* (Courage Is the Victor); this ran for a further 60 episodes. In its two halves, the proverb could have stood as a mantra for the time – but how was it interpreted by the listeners of the era?

[1] Interview with Khaba Mkhize, 1999.

On Thursday 13 March 2014, a large portion of the popular *Vuka Mzansi* (Wake Up the South) Breakfast Show on Ukhozi FM, the post-1990 successor to Radio Zulu, focused on tributes to and memories of Muvo Maphalala, 'Mr Brown', a much loved former broadcaster on Radio Zulu and Ukhozi FM who had passed away two days earlier. Former colleagues and listeners who called in remembered his jokes, his humour, his snappy clothes. Some recalled the music he played. Because it was radio, and the archives were good, listeners could briefly hear his voice again, introducing 'his' music. They heard him leading into a Boom Shaka rap song from the 1990s, 'It's About Time', and then a fast Chico number made for dancing. The formidable Ukhozi FM radio personality Vicky Masuku recalled how he had impressed on her, as a colleague, the need for 'order' and good preparation in the life of an announcer. Sipho Mbatha, 'Sgqemeza' (Big Head), *Vuka Mzansi* Breakfast Show's latest anchor, led by saying how Maphalala always wanted to 'build people up' and played a snatch of 'one of his favourite tunes', 'Georgia On My Mind'. I think it was the Ray Charles version.

As the tributes and memories came in, V. V. O. Mkhize, author and former announcer, listed the names of other loved broadcasters who had 'gone before': Dalton Ngwenya, 'S'jula' Somkhonto, Thetha Masombuka, K. E. Masinga. 'He'll be joining them,' he said. In the midst of this stream of praise and intimate recollection, Mkhize referred in passing to Maphalala as a man 'who made radio plays' (*Ubeqoph' imidlalo*), and in the short calls that came in from listeners, sandwiched between tributes from Advocate Senzo Khumalo and the KwaZulu-Natal Premier Senzo Mchunu, a woman's voice said breathlessly, '*Yiz' Uvalo* – that was *his* drama and it was amazing!'

Muvo Maphalala was indeed the producer of the double radio drama that ran for 100 episodes from late 1986 to mid-1987, and which is remembered by the title of its first part, *Yiz' Uvalo* (In Spite of Fear) and sometimes simply by the name of its chief character, *Sigidi* (Million) (Sibiya 2001b). It was a story of sorcery, of a family broken apart, and of a longing for wealth and a better life. It riveted Radio Zulu listeners for almost six months in 1986–7, in the midst of violence and death, with the country in a state of emergency declared by President P. W. Botha, a low-level civil war in Natal, and bloody disputes between hostel workers and township dwellers on the Rand. The serial drama, produced from the Durban studio of the SABC in what was then Old Fort Road, aired twice a day so that people could find a second chance to listen if they missed the first at 3.05 p.m. The drama maintained a listening community, a loyal public at a time of turmoil. In this chapter, I ask why that was the case and offer some explanations.

Central to understanding the 'pull' of *Yiz' Uvalo* is the nature of radio itself and its ability to create a 'sonic community' not easily definable in terms of ethnicity, region, class, or generation. It was fiercely 'of the moment' – in particular, the nightmare life of what was then the Natal Midlands (and adjacent fragments of KwaZulu), now in the province of KwaZulu-Natal. The drama's magnetic attraction was linked to the power of the unfolding story, but also to the traumatic, 'bewitched' spirit of those times. Within the story and the times was the dangerous and ambiguous presence of the occult, the power of the supernatural. That, centred on the unravelling of a family, was the source of its huge appeal.

The outpouring of loyalty, affection, and fragments of shared memory that played on Ukhozi FM two days after the passing of Muvo Maphalala, who had produced *Yiz' Uvalo* almost 30 years earlier, showed how radio can evoke the memory of an individual, of an era, of a bundle of ideas and physical sensations, through the cues of sound. A man's voice – in this case that of Muvo Maphalala – with its very particular timbre and intonations could call up a listening moment from years before; a cache of music linked to him could connect histories and eras – the civil rights movement and the apartheid shadowlands, and the fast and upbeat South Africa of the 1990s too. Thus it could set the listener back within a stream of simultaneous 'presentness' and recall.

The tribute to Maphalala on the normally fast-moving *Vuka Mzansi* Breakfast Show gave a clue to how the radio station had built an identity and a listening community that spanned decades as well as political eras, moving from apartheid into the 'new dispensation' (from the often used Zulu phrase *uhlel' olusha*), with a listenership of 7.7 million at the time, a figure mentioned on the show in March 2014. The brotherhood of names intoned by V. V. O. Mkhize on 13 March suggested both a family and a close-knit community, with listeners and broadcasters bound to each other through a variety of programmes that had kept people loyal to the station through a time of huge pressure on Radio Zulu from the apartheid state. Radio Zulu – and before that Radio Bantu – was predicated by its practitioners, if not its white managers, on building a community not only of resistance but of the imagination and the senses, in particular the auditory, a kind of aural but also a moral economy of shared understanding of a past and a present. Radio, and the radio dramas, had been able to give voice to all this. There was a community of listeners who interacted with the radio personalities who created programmes and shaped the identities and modern consciousness of listeners as they forged a place for themselves in the difficult decades from the 1960s to the end of the 1980s. It was affectionate, but also demanding. The remembering of Muvo Maphalala and the marking of his death – to be followed by a memorial service the next day and his widely announced funeral at the weekend – recalled the closeness of a former era at the station but also let it be known that, in a very different media world, some of that closeness remained.

The tight-knit community of listeners – based mainly within what was then Natal and the 'Bantustan' of KwaZulu, as well as the areas around Johannesburg – was the world into which the drama *Yiz' Uvalo* stepped, in Radio Zulu's mid-afternoon '*umdlalo womoya*' (drama of the air) slot, on 15 December 1986. It was broadcast at a time of extreme insecurity, of social turbulence, political violence, and psychic angst, but it was also a time of a new incipient consumerism, hence the telling name of the main character, Sigidi (Million). All these factors had a bearing on its reception and on its status as one of the lengthiest and best remembered serial dramas in the long life of the station, rivalled only, perhaps, by the much-loved and idyllic *Isikhumba Sebhubesi* (The Lion's Skin), written by K. E. Masinga's prolific colleague Hubert Sishi, who was also based at the Durban SABC.

The 100-episode serial born into a time of turmoil seemed to set up some consonance between the violence within the drama and the violence without, in the world. The drama forced its listeners to contemplate issues of power and violence and the exploitation of occult power, at a time when lives were being broken, families destroyed, and the use of *muthi* (protective medicine) was openly brandished as invincible. It refracted the outer violence through the highly sonic narrative of the air. Besides the aural economy of *Yiz' Uvalo*, which allowed for a private–public mediation of the feared occult, with its weighty but ambivalent proverb title the drama captured the feeling of the times, the political and social violence.[2] The province now known as KwaZulu-Natal was in some ways a focal point of this, although wider media coverage, particularly international, tended to focus on Johannesburg and the Rand.

The detailed records of John Aitchison and others indicate that political killings in Pietermaritzburg, and in the surrounding areas to the north of Durban, escalated from early 1987 to 1989 (Aitchison 2010a; 2010b; 2010c), but the run-up to this crescendo of violence was itself terrible. In a paper titled 'The civil war in Natal' (first produced in 1989), Aitchison shows how labour conflicts, specifically a strike in the small town of Howick and a consumer boycott that followed, intensified conflict and brutal killings:

> The finale was the bussing in of a large Inkatha group into Mpophomeni township on the evening of December 1986 [sic] that led in the early hours of the morning to the execution of three COSATU supporters. The inquest findings delivered in March 1988 found that nine Inkatha members were responsible for the murders (against whom there have as yet been no prosecutions). (Aitchison 2010b: 80)

[2] The section on the violence in Clermont draws heavily on Wiseman Masango's personal experience of the period. My deep thanks to him.

The end of 1986 and the first half of 1987 built up to 'a horrific period of violence in the Natal Midlands' (Aitchison 2010c: 141). Thus, in February 1987 (when the long second section of the drama was already under way):

> An Inkatha branch chairman was shot and a number of youths belonging to the Hammarsdale Youth Congress (HAYCO), a UDF affiliate, were harassed, abducted and murdered by Inkatha vigilantes. This led to an exodus of youths from Mpumalanga and considerable unrest in the schools. (Aitchison 2010c: 140)

Alongside the painstaking recording of the violence by Aitchison and his colleagues, we can set the memories of my research assistant, Wiseman Masango. He lived in Hammarsdale as a schoolboy but fled to Clermont in Durban, to escape the escalating conflict. Together, the accounts enable us to see the 'time' of the drama. Masango's recollections help construct a picture from the perspective of 'local knowledge' and give a view of the experience of the violence on the ground. They provide a personal insight into the reliance by many involved on the powers of occult substances, the violence of the everyday over and above the structural violence of late apartheid, and, above all, the devastation of family life. At this time, private and public spaces existed in a context of flux and fluidity as violence and counter-violence spread in the region.

Family and politics merged, and to be eminent even through marriage was dangerous. Hammarsdale, the industrial and residential area between Durban and Pietermaritzburg, was affected by violence immediately after the assassination of Victoria Mxenge, the lawyer wife of Griffiths Mxenge, in Umlazi, Durban, in August 1985; both were prominent members of the pro-liberation United Democratic Front (UDF). In the crowded freehold area of Clermont in west Durban, one of the events that sparked violence was the assassination in 1988 of Pearl Tshabalala, the wife of the UDF leader Vuka Tshabalala. Powerful men were often targeted through their wives, hence the Tshabalala murder. Also targeted and killed was Zazi Khuzwayo, wife of an influential businessman in Clermont who opposed the influence of KwaZulu politicians in the area. In Hammarsdale, even for ordinary people far from positions of wealth or leadership, no one felt safe; if a member of your family was an activist (either for the UDF or for Inkatha), you were vulnerable and liable to be attacked and killed or to have your house burned. Thus, the family, already deeply affected by the long-term ravages of apartheid, was newly vulnerable in the urban and peri-urban areas where violence reigned.

In these times of crisis, resort to superhuman powers was widespread. In some cases, the power of the *inyanga* (doctor) and of the politician resided in the same individual: Jamile Mlotshwa, for instance, was an *inyanga* and president of the KwaZulu and Natal Traditional Healers Association. He was also a senior executive figure of the political party Inkatha and was from Clermont. Jamile had a *muthi* chemist shop in KwaDabeka hostel in Clermont. He also had his own *amabutho* (regiments) made up of young men aged from 18 to 25 who were responsible for his safety and guarded him at all times. Jamile, politician and *inyanga*, was eventually charged and convicted of the murder of Zazi Khuzwayo and spent several years in jail before being granted parole.

The intense danger of the times heightened the belief in *muthi* and its efficacy. Moreover, the warlords were particularly careful in their use of it and encouraged its use among their informal regiments. Certainly, in Clermont there was a strong belief in *muthi*; this was fostered by warlord figures such as Shabalala from Emathendeni in Inanda (on the northern outskirts of Durban). It was used by Sipho Mlaba in Hammarsdale and by David Ntombela in the peri-urban/rural districts clustered around Elandskop, north-west of Pietermaritzburg. All these were powerful people who used *muthi* – and they, and others, believed that they used it with success.

Nor was *muthi* only used in Natal and KwaZulu; it was also used on the Rand, particularly in the mine compounds and hostels. An August 1986 issue of *City Press* announced that, in an attempt to find the murderers of two white policemen after a mass meeting of mineworkers, police had sealed off part of the relevant township and searched every house, looking for razor marks used by *izinyanga* for medicine that would turn bullets to water. This belief had a long tradition over a wide region and was also used in the Maji Maji revolt against the Germans in Tanganyika in 1905.

The Rand, and in particular Soweto, was also the site of violent rent boycotts at this time. In a spectacular front-page headline that scooped up American popular culture and the politics of resistance into a neat whole, an article in *City Press* likened the scene in White City Jabavu to being 'like living in the spectre of "The Guns of Navarone"'.[3] At this time, as residents resisted attempts by the municipal police, or 'black-jacks', to evict them, the death toll rose to more than 20, with 'the number of injured galloping towards the 100 mark'.[4]

All this was taking place in the context of widespread detentions under the Internal Security Act, with 'over 460 detainees' hospitalised between mid-June and the end of August 1986.[5] There were also continuing executions by the beleaguered apartheid state, including that of Andrew Zondo, 'the Amanzimtoti bomber'.[6] This was the spectacle of public and private violence that the serial drama *Yiz' Uvalo' Umanqob' Isibindi*

[3] *City Press*, 31 August 1986. [4] Ibid. [5] *City Press*, 14 September 1986. [6] Ibid.

mediated through its script, its eerie sound effects, Tina Turner's 'We Don't Need Another Hero', and the voices of its glittering and much loved cast of players.

Textures of production and reception

Certainly, the *imidlalo yomoya*, or radio serial dramas, were still a key creative feature in the life of Radio Zulu in the mid-1980s. Letters from listeners were usually sent in at the end of a series and could be full of praise or fiercely critical. Eric Ngcobo, who played the lead role of Sigidi in *Yiz' Uvalo*, recalled that letters 'would come at the end [of a drama] and say things like: "It would be better if So-and-so hadn't died." People would write from all over.'[7] The letters written at the end of *Yiz' Uvalo* have not survived; their comments, although interesting and important as a gauge of the audience's response, would have been unlikely to change Morris Bhengu's plans for the second part. The first and second parts followed on with only three days in between, so it is likely that Bhengu, the author, at least began *Umanqob' Isibindi*, the 60-episode second half, well before the end of Part 1.

With most of the dramas, the expectation of listeners was that they would contain some sort of message to help them live. *Yiz' Uvalo* was unusual in the ambiguity of its message. Finally, there was a resolution of sorts, and the powerful and destructive *inyanga* Gonondo dismantled his remote homestead and returned to the even more remote 'Tanganyika'. Yet he was not caught by the police, who found only ashes and heard a laughing voice as they entered the last house. Sigidi, the suffering father who became a mass murderer, was punished by the loss of his family and had to lead the life of an outcast wanderer, but he was never arrested. There was an ambiguity about the potential power of the supernatural in daily life which unsettled some listeners, an uneasiness that some interviewees recalled when they spoke about the play over a decade after it was broadcast.[8]

The angst some listeners felt in relation to the play, and their equal fascination with it, reflected an important element that was part of the modernising of people's lives. Beyond the political violence, the 1980s was an era of tumultuous change in material and social terms, but that did not mean that magical powers faded away. Adam Ashforth's account of life in Soweto in the 1990s notes the lengths to which people went to defend themselves 'against the possibility of harm from substances imbued with magical power' (2005: xv). He writes, too, of the general

[7] Interview with Eric Ngcobo, 2011. See also Gunner (2000a), Gqibitole (2002), and Sibiya (2001b).
[8] Interviews with Wiseman Masango, 1999 and 2004.

condition of spiritual insecurity and of 'learning of comrades who had been murdered for their body parts to be used in magic medicines' (ibid.: xv). He notes that '*muthi*' was the central term in the discourse of power linking human action to invisible agencies, and later refers to the term '*ukuthwala*', which he glosses as 'using the occult force that is believed to be mobilized through *muthi* to gain wealth or power at other people's expense' (ibid.: 2).

Such anxieties and practices have a long history in the region, and far more widely.[9] The context of the desire to control and change the course of events through the use of magic medicines made from body parts was discussed by Harriet Ngubane in her 1980s study of sorcery and 'ritual murder' in Swaziland (1986: 194, 203).[10] Ngubane stressed that such practices, tied to specialist knowledge, were 'ritual', not simply wanton killing, and were linked to crises that often related to chiefly power. In a similar vein, Rob Turrell's study of court cases of ritual murder in early colonial Natal noted that, by the 1920s, 'the resort to *muti* [sic] murder "was no longer, in Harriet Ngubane's terms, an indication that something had gone seriously wrong"' (2001: 37, quoting Ngubane 1986: 203). What seemed at one point to have been a 'right' of rulers to commit ritual murder when access to power was urgently needed, often in times of social and political stress, by the 1920s had become a practice to which commoners and ordinary people had found access and was called, in isiZulu, *ukuthwala.*[11]

Bhengu's serial drama exploited this domain of the metaphysical and supernatural within the modern some 60 years later with a radio-technical magnificence that brought feared and hidden knowledge into the intimate living space of all who listened. The use of the figure of the specialist doctor/healer, the *inyanga*, in radio drama did not begin with Bhengu. On one level, it was a staple device that writers could draw on when they felt they needed to; it was something they had done quite freely in the previous two decades. But it had to be handled carefully, and scriptwriters had different approaches to its use. A decade or so after the play aired, when discussing what they called 'the *inyanga* factor', Bheki Msane and Dumisani Nkosi, both writers and producers of radio drama, took opposing positions. Msane maintained that using a herbalist spiced up your story: 'As it is a kind of taboo: people, "modern" people, are not supposed to go to herbalists any more, but they do, and it spices up a story to talk about taboos'.[12] As Karin Barber has remarked in relation to Yoruba popular theatre:

[9] See also Agamben (1993: 12) on sovereign power and bare life.

[10] Cited in Turrell (2001: 21–38).

[11] The word has many meanings in isiZulu, the most common of which is simply 'to carry a load'. In isiXhosa, however, it can mean 'to abduct a bride'.

[12] Interview with Bheki Msane and Dumisani Nkosi, 1998. Note: this point comes from Gunner (2000a: 48).

> Modernity meant not the rejection of the traditional but the possibility of selectively recuperating it – and the editorial operation by which this was accomplished was subject to continual redefinition over the theatre company's lifespan. (Barber 2000: 424)

The continual redefinition in Zulu radio drama of the difficult and ambiguous areas of 'tradition' certainly applied to the *inyanga* figure. Dumisani Nkosi's position differed from his colleague's: he tended to portray them as liars (see Chapter 7). For Nkosi, the melodramatic was part of his formula, but what really mattered was a fidelity to what happened in people's daily lives. The recurrent appearance of the *inyanga* in the dramas of the 1960s and 1970s would suggest that the figure was indeed often within the schema of everyday life. As we saw in Chapter 5, the ambiguous knowledge and power of the *inyanga* featured in the mid-1970s classic love story *Ubongilinda Mzikayifani* (You Must Wait for Me, Mzikayifani), which was re-broadcast in 2000. There, it showed that an *inyanga* could heal illness and also carry out devastating retribution in the form of a fatal storm directed at the home of the man who had stolen his precious cache of medical knowledge. It was this ambivalent feature of indigenous power and knowledge that the drama *Yiz' Uvalo* exploited, as the country waded through blood, struggling towards the new era.

By the mid-1980s, the radio dramas on the station known since 1973 as Radio Zulu could draw on a stable of writers, actors, and producers that would have been the envy of any radio station in the world (SABC n.d.: 76). As had been the case since the station's earliest days, its producers still doubled as actors and announcers and were often well known as the leading voice of another programme. Sagius Thwala, for instance, who ran the philosophical Sunday programme *Idwala la Phakade* (Rock of Ages), was also an actor/producer, and there were others.[13] This multitasking was perhaps a generic feature of radio production (Hendy 2000). Yet, in the case of Radio Zulu, the doubling up of roles and skills was linked to limited resources in an SABC station that was under surveillance and lacked the funding of the English and Afrikaans services. However, its practitioners, who were determined to go beyond the servile, mouthpiece role ordained for the station, had a restless vision of what was and what might be possible. This had driven them at least since the 1960s, with versatile masters of sound such as Alexius Buthelezi, Hubert Sishi, and the great K. E. Masinga himself.

In the view of some, including the novelist and scriptwriter Nakanjani Sibiya, the 1970s was the high point of radio drama.[14] But the cast of *Yiz' Uvalo*, the dark jewel of the 1980s, contained some illustrious names. Lindiwe Ntuli (playing the wife of Sigidi) and Lawrentia Madlala

[13] Interview with Eric Ngcobo, 2011. [14] Interview with Nakanjini Sibiya, 2011.

Figure 6.1 Lindiwe 'MaLindi' Ntuli, one of Radio Zulu's and Ukhozi FM's best-known and loved radio drama voices. Possibly her most famous role was as MaMbatha, the wife of Sigidi, in *Yiz' Uvalo*. *Source*: *Ilanga* Photographic Library.

(Sigidi's mother) were two of the stars of the drama, central to its core family aspect. Other well-known names such as Douglas Mfeka and Shadrack Ngema were in the cast. Eric Ngcobo as Sigidi was matched in acting skills by Welcome Bhodloza Nzimande, who played the fearsome *inyanga*, Gonondo, 'On whom the eyes never gazed [*Ongabhekwa ngamehlo*]' (see Sibiya 2001a). The part of the *inyanga*'s assistant, Makhas' Omdloti (Tobacco Leaf), was played by Lawrence Sithole. His role made such an impression on the listening public that it dominated his obituary in 2002, a decade and a half after the series' end.[15] Also, in a minor female part was the young Zamambo Mkhize, who would go on to be a manager at Radio Zulu, and then later programme manager on SABC 1 TV (SABC n.d.: 112–13).

The radio personalities who appeared in the long drama had their own distinctive listening 'pull'. Listeners imagined them on two levels: first, as the character in the drama, moving through its extraordinary and tragic events and momentarily transformed into the figure within the drama;

[15] *Ilanga*, 22–24 August 2002.

Figure 6.2 Muzi Eric Ngcobo, prolific radio dramatist for Radio Zulu and Ukhozi FM and an actor of renown. He was MacDuff in Welcome Msom's stage production, *uMabatha*, and loving husband turned murderer in Morris V. Bhengu's *Yiz' Uvalo*.
Source: *Ilanga* Photographic Library.

and second, they knew them as personalities in the world of radio who played many other roles, both dramatic and as radio announcers or presenters. Each character/individual was thus identified as a complex persona within the broader domain of popular culture, which the radio programmes brought into being and continuously remade.

The station was under-resourced, and its public – by now very well formed and with the confidence of its own particular voice – was one that was partly obscured from view in relation to the 'official' apartheid culture. Yet its cultural resources were huge. There were the multiskilled artists of the radio serial scripts – the producers, actors, and writers – and, at a little distance and invariably white, the sound engineers and technicians. Vic Dralle, known by the name of Unogwatsha (Rabbit), was the sound technician for *Yiz' Uvalo*. Eric Ngcobo considered him 'the best in the country'. As the character Sigidi, Ngcobo was particularly aware of Dralle's innovative sonic inventions evoking the paranormal sensory nature of magic – the eerie whistles, echoing voices, muffled drumbeats, and so on – that were all indispensable to the terrible metamorphosis of the character from family man to killer.[16]

[16] Interview with Eric Ngcobo, 2011.

In 1971, 15 years before Eric Ngcobo took on the role of Sigidi, the frustrated family man who turns to occult powers to fulfil his dreams of wealth and well-being for his family and himself, he had played MacDuff in the London *uMabatha*. *UMabatha* (the Zulu version of *Macbeth*, adapted by Welcome Msomi) and *Yiz' Uvalo* had some points in common. The focus in *uMabatha* on ambition and on apparent success secured through the use of powerful charms resonated with the famous radio drama, which had magic, ambition, and violence at its centre. What the radio serial captured for its late twentieth-century listenership was the very modern nature of the occult and the place it had in the life of struggle and insecurity in which its central protagonists, particularly Sigidi, were involved. Moreover, as has been cogently pointed out, the occult is within the experience of the modern, not outside it (Meyer and Pels 2003).

The mix of the drama

In an interview in 2000, when Eric Ngcobo looked back on the famous serial drama for which he had won a coveted SABC Artes Award for his acting, he made it clear that it had almost subsumed him. It was a drama that pursued him for years after the serial was over. His real name and personality became submerged in those of the character he played in the daily, Monday-to-Friday drama over the six months from December 1986 to May 1987. He recalled that for years, as he walked down the streets of Lamontville in Durban, children would run screaming at the sight of him, calling out 'Sigidi!', imagining that he would kill them for *muthi*.[17] Looking back during a second interview in 2011, he had his own theory of the origin of Morris Bhengu's ideas for the story. For him, it was nothing to do with 'warlords'. Firstly, he saw it as a study of a mass murderer, the first of a series of dark dramas that Bhengu wrote for radio between 1986 and 1992.[18] He then elaborated on the man who may have been Bhengu's model for Sigidi: in the mid-1950s, Ngcobo recalled, a man named Ngqavini Msomi was convicted for killing a number of women. He would murder them, then cut out their private parts and make bangles with them, which he would sell. This occurred in the area south of Durban. The case became notorious, said Ngcobo:

> and [it] was in all the papers – *UmAfrika*, *Ilanga*, the *Mercury*, the *Daily News*. He would get caught and would miraculously escape: a rat or a cat would rush out, [but] it was him!! But they got him in the end. And that's where he got the concepts from. It was *Big News*![19]

[17] Interview with Eric Ngcobo, 2000.
[18] See, for instance, *Bakikiza Buthule* (1999).
[19] Interview with Eric Ngcobo, 2011.

At this point, such suppositions cannot be verified. Yet they illuminate how the drama series grew out of a combination of 'the known' and the bizarre everyday events in life in which the occult – the 'rat' escaping, the 'cat' eluding its captors – had a place. Certainly, the drama series was dark, with the struggle of a family to survive as its central reference point. In his effort to provide materially for his family, Sigidi destroys it, and himself. A key motif linked to the family is the role of the ancestors: that is, the paternal ancestors of the protagonist. The paternal ancestral names, ritually called upon as early as Episode 3, provide an important link with the supernatural, and with the esoteric knowledge and dangerous occult power that permeate the play.

Another theme that expands as the drama unfolds concerns the loss of family and personal order, of calm, and the onset of massive instability. This is shown in various ways: the protagonist, back in rural Ndwedwe, north of Durban, after back-breaking work in 'eGoli' (Johannesburg), wishes to make a new start for his family. He achieves it briefly by selling fruit and vegetables to local people from a small van, bought on credit from the local garage. But the obstacles pile up. He becomes increasingly isolated from normal human ties (which the drama repeatedly evokes) and bound more and more to the world of the occult.

The theme of the world beyond the material realm is always present in the series. It is evoked through particular sonic devices such as drums, which signal the presence of the otherworldly and a transition to a different state of mind and being. The drums enter quite early in the drama, first in Episode 12, when Sigidi, who is in search of good luck (*inhlanhla*), first meets the *inyanga* in his remote dwelling at KwaMhlabuyalingana, in the north of today's KwaZulu-Natal, near the border with Mozambique. The drums return with heightened insistence in Episode 13, to mark the entry of the protagonist on his fatal path. Behind a curtain of skins, his voice the only sign of his presence, sits Gonondo Nyathi, feared dispenser of occult power, who will change Sigidi's life forever. Also laid out for popular consumption as Sigidi encounters Gonondo, and later at other strategic points, are gobbets of esoteric knowledge. In this first meeting, these are linked to the strange animals that represent the power of the *inyanga*: the lion, the huge snake, the crocodile. They represent the forces that Sigidi has to be strong enough to master if he is to continue on his path to 'success'. Other details of such knowledge are introduced at later stages in the drama as the listeners follow Sigidi on his terrible journey.

Not a world apart: the plot of *Yiz' Uvalo, Umanqob' Isibindi*

The plot of *Yiz' Uvalo* (used as shorthand for the double series) made it part family drama, part murder mystery, and part examination of the intimate and fearful occult. Its events and predicaments seemed outrageously far from ordinary experience, yet closely linked to it as well. The episodes began on the familiar note of the family and the domestic, with a man, Sigidi, ready to earn a living for his family. His wife, MaMbatha, mother, MaNgidi, and his two children, Nomzamo and the teenager Phakamani, are introduced early on. All seems calm and hopeful, but early success is soon followed by problems. The local storeowner, Nkabinde, objects as his customers disappear overnight and buy cheaper and fresher products at lower prices from the mobile newcomer, Sigidi. The shift from domestic harmony and brief financial security to the violent and the unstable soon follows. By Episode 10 (thus at the end of the second week of the daily drama), Sigidi, in despair at his failing business and rising debts, has decided to visit the 'Thonga man, more powerful than Khotso', called Gonondo Nyathi. The direct comparison with Khotso (Sethuntsa), the widely feared, fabulously wealthy, and powerful magician from the Transkei who died in 1972, would have signalled to listeners that the character they were about to encounter over the airwaves was no ordinary *inyanga* (Wood 2003).

The turn to the occult and to hidden knowledge (*ulwazi olufihlekele*) is carefully introduced in the following episodes. Soon, the supernatural and the domestic feature together in Episode 11, when Sigidi begins to fall apart psychologically and turns violently on his family. Bridging music enables the swift cuts between moods and locales. After this point, as Sigidi finds himself in the new domain of occult power of *muthi*, the domestic and the humane seem to disappear from his personality. The close focus through Episodes 13 to 17 on Sigidi's ordeals shows the protagonist surrendering himself to cruel forces greater than himself in his quest for wealth and power beyond the human. He has to kill a woman who has been 'troubling' Gonondo as part of his test; as he does so, the call '*Mthathe Sigidi*' comes from his master and signals the loss of his old self.

The dehumanisation continues through a perversion of language as humans become 'two-legged goats' (Episode 17), and details of killing become the new domestic vernacular. There is no escape for Sigidi, either physically or psychologically. When he returns 'home', his wife discovers a human finger in his bag. His unstable violence, which appears between brief, contrasting flashes of domestic happiness (Episode 22), leads him to murder two more people in quick succession – the garage owner and storekeeper who had thwarted him earlier, and then a random young woman to whom he gives a lift.

The normal and the paranormal are entangled as these violent events play out and the bridging music allows shifts of location. A development in the plot comes when one of the policemen, Simamane, who is looking into the murders, experiences the tactile presence of unseen objects and decides to leave the force and begin his own undercover investigations.

He traces the killer's path with careful logic and forensic skill, but later he too becomes ensnared in the coils of the powerful occult. As the main plot moves forward, the ability of sorcery to take over the mind of the victim (Ashforth 2005) takes a new turn as the message to kill is increasingly internalised by Sigidi. A fifth 'two-legged goat' (a young woman in a bar) dies and her friend narrowly escapes, but they are only a prelude to Sigidi's increasing desire to kill his wife. *She* is to be his sixth 'two-legged goat'. The mantra '*Mthathe Sigidi* [Take her, Sigidi]!' dominates, as Sigidi struggles with killing his wife, even at this late stage of his possession.

The tension of and listener interest in these last episodes of *Yiz' Uvalo* – which by now was nearing its fortieth, and final, episode of Part 1 – must have been great, as two parallel but linked plotlines reached their climax. The clever policeman, attempting to unmask the criminal at his source, poses as a would-be client of the *inyanga* Gonondo, and thus plays an extremely dangerous game, as listeners knew only too well. At the same time, the final killing of Sigidi's wife takes place – or so it seems.

As the 40 episodes drew to a close, the themes of the family as a site of possible harmony and of the police as potential bringers of order in a disturbed world sat alongside the theme of occult power. Episode 40 closed on what appeared to be a perfect resolution, and a restoration of justice. The ex-inspector Simamane, now undercover and known by his new name of Makhekhe (Biscuits), 'to sweeten the violence', has courageously captured not only the *inyanga* Gonondo but also his accomplice Makhas' Omdloti and family man turned killer Sigidi himself, and has returned them all to an amazed and delighted Captain Ndlovu in Ndwedwe. Thus, at the end of the 40-episode drama, which ran non-stop five days a week (including Christmas Day), there was a resolution and order had been restored after extraordinary suffering and violent destabilisation – and huge engagement on the part of listeners. In the world beyond the radio drama, as this section ended in February 1987, violence and mayhem reigned, with the Natal region one of its epicentres and the Rand another.

Only three days passed between the apparent resolution of Episode 40 and the airing of Episode 1 of the new serial drama, whose title completed the proverb, grounding it in the idea of courage as the quality that could overcome all else: *Umanqob' Isibindi* (Courage Is the Victor). It began on 9 February 1987 and ran until 2 May, playing twice daily on weekdays. During this time, violence – political, social, and personal – continued to wrack not only the Natal region but also the Rand (Bonnin 2000; Bonner and Ndima 2008). Listeners to Episode 1 of the sequel would have learned to their surprise, shock, or delight that the happy ending of three days before was only an intricate dream played out in the undercover policeman Makhekhe's subconscious. Listeners who may have hoped for some sort of clear message about how to lead a good life despite great trials would hear nothing of the sort. The words 'The victor is courage' come from the mouth of the *inyanga* Gonondo as he praises Sigidi for his endurance. But even Gonondo is equivocal as he urges Sigidi to murder his wife. 'Whatever you do now, you must do it for the children,' the *inyanga* counsels. He continues: 'Remember that in whatever you do, "*Umanqob' Isibindi*".'

In the 60 episodes of *Umanqob' Isibindi*, which carried listeners through to a strange ending, there is a slide into a quickening violence, with many things unravelling. Yet the reduced family, embodied by Sigidi's despairing children, remains one of the abiding themes. The fate of mothers, wives, and grandmothers as key units of the family mattered greatly to listeners. Thus, Sigidi's injured but not dead wife, MaMbatha (Lindiwe Ntuli), continued to play a role, as did Granny MaNgidi (Lawrentia Madlala). The focus in the tense and fast-moving sequel was on how people are transformed from their former selves. It is the metamorphosis from the normal to unrelieved destructive capacities in a single individual, Sigidi, that gathers momentum in Part 2. And yet it seems that this process is never quite complete: Sigidi struggles to keep track of his children and to send them money, even when it is quite clear that he is lost to them forever as a father. Simamane, the rogue undercover policeman who becomes corrupted, cannot escape his transformation into a debased human being, forced to kill for the *inyanga* Gonondo. Yet he, too, fights against total transformation and keeps in contact with his mother to the very end. Taken together, the episodes show a gradual but persistent degradation: Gonondo refers to people to be killed as 'garbage' (Episode 30); he counts people killed in terms of cash, and remarks at one point that the killing of a woman is more highly paid than that of a man because it is more difficult.

However, in the grim episodes of the sequel, the family fights back. In spite of the loss of the family as a site of harmony and normality, the children still try to preserve what was once a working family, even as another murder of someone they know is reported, the corpse's eyes having been removed. In the last episodes, as the unravelling of normality continues, Sigidi, still alive, is a wanderer in the world, outside the social order. Briefly finding a new identity in a small community called Ngagane (Thorny Place), he begins a new, 'clean' existence, but this doesn't last long. The life of the freelance undercover detective Simamane/Makhekhe, too, is changed forever.

Sigidi, by now as good as a 'ghost' and deserted by his ancestors, has no home except that of his *inyanga* master, Gonondo. As the second series of 60 episodes drew to a close, Simamane/Makhekhe, dying in hospital after a failed bank robbery (having forgotten his protective *muthi*), in a moment of melodrama reminiscent of the popular travelling plays of Gibson Kente (Peterson 1994), writes a final, moving letter to his mother. At the very last minute, in the final episode, there is some sense of evil on the run. Gonondo orders his accomplice Makhas' Omdloti to return home, as he begins to dismantle his establishment. The latter realises that perhaps Gonondo was never a real human being, but rather a dead but living person, *umungcwi*, who now wishes to return to his distant maker, Mkhandwiri, in Tanganyika. Gonondo's final words to his two underlings again use the proverb that links the two dramas together: *Yiz' Uvalo, Umanqob' Isibindi* (In Spite of Fear, the Victor Is Courage). And yet that is not the very end. In the final minutes, in Episode 60, as the helicopters stutter above Gonondo's homestead and heavily armed police encircle it, the policeman who breaks into his home finds in the last rondavel only a heap of bones. A laughing voice addresses him:

Wephuzile Mbokazi. Ayikho into ongabe usangenza yona. Usuyozwa ngendaba ukuthi wayenza imisebenzi mini uGonondo. [Ahleke kakhulu] Ngeke usangenza lutho Mbokazi, Umanqob' Isibindi! [Ahleke kakhulu]

You're too late, Mbokazi. There's absolutely nothing you can do to me. You'll hear in due course the story of Gonondo's deeds. [Loud and long laughter] You can't do a thing now, Mbokazi. Courage is the victor! [Loud and long laughter]

Balanced against this, however, just seconds earlier, with mood and scene switching with all the ease of radio, listeners would have heard Simamane/Makhekhe, the rotten policeman, confess everything to Captain Ndlovu as he lies dying in his hospital bed. There follows (again reminiscent of Gibson Kente dramas) a rich, flowing gospel song sung by the police, with the words:

Amagugu alelizwe / Ayosala emathuneni.

The country's precious ones / Will be left in their graves.

This glorious melody fills the soundscape just seconds before Gonondo's final, eerie celebration.

It was only in the late 1990s, after Wiseman Masango had conducted over 20 interviews with radio listeners on the Pietermaritzburg campus of the University of KwaZulu-Natal and in Peacevale, Clermont, and Hammarsdale (all near Pietermaritzburg), that the importance of the double series *Yiz' Uvalo* in people's memory over years of listening to radio dramas became clear to us (Gunner 2000a).[20] The people who were interviewed seemed to remember and value most the sensory mix of the drama: the 'play' of the language, the tensions, the interactions of characters. They also valued the resolution and the way in which forms of evil

[20] Wiseman Masango: interviews with 22 listeners, Pietermaritzburg, April 1999; interviews with 4 listeners, Peacevale, October 2004.

were worked through to a positive and moral resolution – which was how they thought drama ended.

This was when we decided to find out more about the secret of *Yiz' Uvalo*'s power to live on in the social memory of ordinary people. Wiseman Masango went to the SABC studios in Durban, where, thanks to the generosity of Selby Goba, then drama archivist, and to the station manager at the time, Khaba Mkhize, he spent many hours closeted in a studio, listening to tapes of the play. There were no extant scripts that we knew of. Sometimes he made summaries, at other times transcriptions; he worked in isiZulu and in English in a fairly arbitrary manner. We would then discuss the 'script' as we had it, and attempt to unpack its multi-layered hold on its listenership, which was not transient but was still with us.

Masango's skills as a first-language speaker of isiZulu and as a lover of radio dramas interacted with my far less fluent and far more pedantic language skills and my fascination with the dramas as an 'established sign' within popular culture and social life, a crucial part of the mediatisation of voice within the overall technology of radio sound. This was when we interviewed Eric Ngcobo (Sigidi) and heard about his intense memories of the play and its lasting impact on his life and his public persona. Yet its link with the violence of the times, commented on by Khaba Mkhize (the SABC Durban station manager), and Mkhize's statement that 'It's about warlords' intrigued and puzzled us.

Sometime later, in a seminar, a student from Imbali, Pietermaritzburg, touched again on the link with violence. He recalled that, during the low-level civil war of the late 1980s in the Natal Midlands, a pale blue combi-taxi with the name *Yiz' Uvalo* painted on it and dark windows would come swooping down from Elandskop, territory of the Inkatha warlord David Ntombela. No one from Imbali or the ANC-loyal areas of Edendale would dare take it, he told us. So, was the drama's place in popular memory – and radio aesthetics – linked to a memory of violence, and, perhaps, to the experience of violence in different forms: either the symbolic violence of the radio drama, with its powers of language, mediatisation, and radio techniques, or the violence of a particular era? Both were rooted in the quotidian, the materiality of the everyday. Both, too, were linked to the occult, which was also part of daily life. Possibly the drama worked through the pain and evil, while consuming its audience's attention, and so enabled a kind of closure through narrative. Like a terrifying oral tale, Zulu *inganekwane*, or Xhosa *intsomi*, it enabled the contemplation of evil and violence, the listeners' fearful and pleasurable absorption in it, and then a return to some kind of normality (Scheub 1975; Canonici 1988; Hofmeyr 1993), and, perhaps, a moral order restored.

Memory replayed

In the early years of the new millennium, with the promise of *ikusasa lethu* (our tomorrow) still optimistically in the air, and with the growth of new media in addition to the allure of the image and the spread of television among many low-income households, what would be the effect of a replay of *Yiz' Uvalo*? Would it still grip its listeners? Would a further set of interviews reveal a different audience pattern of memory and preference?

A re-broadcast of the first part only, the 40-episode *Yiz' Uvalo*, took place in 2004, but it seemed to pass without too much comment. Masango and I were intrigued and wanted to probe further. In late October 2004, Masango conducted a small sample of four interviews, all in Peacevale. He asked far more specific questions than in the bigger sample in 1999, relating them more closely to *Yiz' Uvalo*. Three of the four remembered the first airing of *Yiz' Uvalo* vividly. One linked it specifically to the period of *udlame* (political violence) and remembered the fear of listening in the dark and not being able to go outside because you might be murdered. The same man remembered with horror Sigidi's violence, but beyond that discussed the way in which 'stories' helped you 'build':

> It's his violence I remember; Sigidi's the one I still remember. It terrified me like no other one ... and now if I go back to those old stories, and some of the new ones too, they do give you tips about life, because these days *we don't really know what lies ahead* [*ngoba manje ikusasa alaziwa*].[21]

The same man thought that 'the old stories' were still needed as they showed children that there were witches in the world, that there was social violence and abuse. While this interviewee wanted the old stories back, he liked the '*izitori*' (stories) about AIDS and ones that helped him manage his own love life. In his view, the tough subjects in the new plays enabled people to discuss difficult topics, such as AIDS.

So the later interviews kept a place for 'Sigidi' – the man and the drama – but gave a clear sense of new preoccupations pushing in and becoming part of the debates on public and private life. Two more interviewees, both women, liked plays that tackled violence against women – a theme hidden in the fearsome occult strand of *Yiz' Uvalo* – and also appreciated the importance of the new AIDS theme. In the last interview, an electrician in his mid-thirties discussed a play he remembered hearing as a child about marriage and then suddenly turned to *Yiz' Uvalo*. He recalled his fear of the dark outside because of it and then listed, as if he had listened to it only a few days before, the names of

[21] Italics added.

Gonondo's occult beasts: the lion Bhamuza and the snake Mkhoco. He recalled, too, the detrimental results of using *ukuthwala*, the *muthi* for wealth and riches. 'If you kill, it will come back to haunt you ... tomorrow it will turn against you.' It seemed that the old play, the dark jewel of the 1980s and one of the high points of the vast sweep of Zulu radio drama, still held some of its pull on the memory and the moral imagination in 2004. It may not have still been a hot topic in the taxis and public spaces, but its unsettling power and its ability to make people ponder the limits of ambition and the everyday occult remained.

Perhaps Morris Vusumuzi Bhengu's play – the drama of all who 'made' it and of those who first heard it – has slowly moved to the status of 'the classic', stately, remembered, but not part of the fast flow of the living stream. Its complex mix is still a resource for future makers of drama in a far more challenging media world. David Hendy, commenting on the dramas on BBC's Radio 4, remarked:

> The best among them ... blended the extraordinary with the quotidian and had an ear for the agony and laughter of 'ordinary lives'. If they featured heroic deeds, they were often as not the deeds of the meek. (Hendy 2000: 219)

Yiz' Uvalo did more: it kept the 'deeds of the meek' and their agony. But it kept, too, the horror of the 'warlords'.

7 '*Ikusasa lethu*' (Our tomorrow)
The 'glorious decade'? Radio drama of the 1990s

One story follows another, like life. — Thando Ngubane [1]

Alikh' iqiniso emhlabeni!
There's no truth in the world! — Matha in *Isiboshwa* (The Prisoner)

Probing the 'new era'

As the April 1994 elections came to a peaceful conclusion, the joy and relief at the end of the harsh years of apartheid and the expectation of what was to come were palpable. Hence the phrase that was on the lips of many, which captured this expectation of a better future for all: '*Ikusasa lethu*' – 'Our tomorrow'. It was a phrase that found its way into many of the dramas of the 1990s, as a prolific number of scripts and a new surge of inventiveness sought to probe the 'new era', often referred to in isiZulu as '*Uhlel' olusha*' (the new order), trying to capture its restless, joyous, yet uncertain spirit. Practitioners of the vibrant popular performance genre of *cothoza*[2] in Durban took the phrase for the name of a new team (*iqembu*), and so the group *Uhlel' Olusha* (the New Order), was born. Like the radio dramas of 'the new era', it was ready to work with the grit of daily life, and to make art for their audiences from what lay around them. Similarly, the phrase '*Ikusasa lethu*' held a notion of community. The use of the plural possessive – 'our' (*lethu*), tied to 'tomorrow' by its shared class concord – signified the collective sense of entry into a new era and a future marked by a communal state of 'betterness' that would continue. It was not simply a single 'tomorrow' but rather the start of a better and radically changed and shared future time. In part, it was the long years of the station itself, from its early time slots under Radio Bantu to its life as Radio Zulu, and then the transition to Ukhozi FM, which had helped create the sense of a community made and shared through a host of voices among generations of listeners. The station had given its voice to a new sense of the black modern, in part through the long river of

[1] Interview with Thando Ngubane, 2002.
[2] *Cothoza* is a form of a cappella singing combined with dramatic action and dance moves.

imidlalo yomoya, which were also tied to the multiple roles broadcasters played on other programmes. Identities had been shaped or tested by the experiences created by the dramas. People had lived through history, and the prism of radio sound had been part of the way in which that history had been experienced.

What would happen now to the radio community and the phrase on people's lips, '*ikusasa lethu*'? In some ways, partly due to the internal changes within the African-language stations after 1990, important adjustments were made. Those seen as being too close to the old guard left, and a newer regime of trust began. Much of this occurred in the 1994–7 era, when Govin Reddy was head of radio at the SABC and the African-language stations fell squarely within his remit.[3] In terms of wider changes, however, the expected 'tomorrow' of the 1994 post-apartheid dispensation soon became a social and economic desert of redundancies, casualisation, endemic poverty, and the rampant and unchecked spread of the AIDS pandemic, which was not recognised or addressed by the state until as late as 2008.

How did this disjuncture between rapture and dismay, the utopian and the dystopic, play out in the serial radio dramas of the late 1990s and early 2000s? How would this seasoned genre, no longer emergent but by now ingrained in the very idea of the radio station itself, play out the life it saw around it, in all its sudden shifts and unexpected directions. It was, after all, not a fantasy genre, but one linked to the struggles of everyday life and the constant expectation and need to craft and imagine a 'good' life, a life that one could also portray as modern. The *imidlalo yomoya* radio genre was one that had the concept of 'the moral' at its heart. At the same time, it was an art form of sound – of the air – with all the symbolic power, specificity, and multivalence which that implies. Possibly what is also important is to remember that, in some ways, what acted through the dramas was a deep 'sedimentation of sound' (Hirschkind 2006) that sat in the memory, the consciousness, and even the listening body.

After 1994, the decade was still close to the euphoria of the country's first democratic elections and the accession to power of the ANC led by President Nelson Mandela. Yet trauma, captured for the country by the testimonies given to the Truth and Reconciliation Commission (TRC), which were aired widely on both television and the stations of the public broadcaster, the SABC, was never far away (Krog 1998; Sanders 2002; Cole 2009). As both Antje Krog and Catherine Cole have pointed out, the TRC hearings were often highly performed events. The testimonies often referenced long-established traditions of oral and dramatic performance, which resonated with the cultural repertoires of the country's

[3] Govin Reddy memorial service, SABC, 23 October 2017. Testimony of Krivani Pillay, SAfm, SABC.

African languages, rather than with English or Afrikaans. Many of the formal accounts of the TRC testimonies, played out in community halls and public spaces up and down the country, were televised and also widely listened to through the medium of radio. Frequently, too, the testimonies were complemented by interviews carried out by particular stations.

On one such occasion in the late 1990s, relayed on what was by then Ukhozi FM, an elderly parent from Pietermaritzburg district gave a detailed account of how he now knew the way in which his daughter had died (she had been one of the victims murdered and her remains then burned at Vlakplaas, one of the notorious 'farms' of the old regime). He recounted the gruesome details of her death, described her remains, and told the Ukhozi FM radio listeners that knowing did not remove the pain but it did provide him with some relief.[4] Such sombre narratives thus existed and were circulated in the public domain alongside the joy of the new. Pain was undoubtedly part of the memory narratives of '*ikusasa lethu*'. Perhaps, however, it was also a time for different stories of the self (Nuttall and Michael 2000), which sprang up together with a new desire to consume and to acquire (Posel 2010). Could these narratives also voice a sense of being a new subject in a new, if conflicted, democratic space?

In such a tumult of national psychic change and shifting social and national imaginaries, what would the radio dramas do, and how would they mediate this context in their complex sound events? Would the genre rise to the occasion? What narratives could possibly capture this joyous but fractured time? In addition, what public did the dramas now serve and what community of the air still claimed the dramas as its own? These are the questions I explore in this chapter, drawing on a selection of the Zulu serial dramas known as *imidlalo yomoya* (plays of the air) or *amastori* (stories) from the late 1990s and the early 2000s. The dramas continued to be aired three times daily on weekdays, thus occupying the same time slot in the new, changed era as they had done previously. What did this temporal sameness mean? In the intimate rhythms of daily life that radio had helped to structure for so long, the *imidlalo yomoya* continued in the new era to mark a node of narrative in the morning, one in the afternoon, and a third in the evening. Two different serials ran in a single day, Monday to Friday, with one repeated on the same day. This had been the pattern for many years, with some small variations. So the appetite for *amastori*, apparent for so long through the intimate listener–station interchange, continued, in spite of the pull of television and the increasing presence of soaps such as *Generations* on the SABC

4 I listened in Pietermaritzburg to the programme on Ukhozi FM but was not able to trace any record of it in the Sound Archives when I and Dennis Mseleku, SABC Durban archivist, searched for the original tapes.

channels. In many cases, these now placed more emphasis on productions in isiZulu and offered a chance for radio actors to move across into the 'new' medium. Some, such as Shadrack Ngema, did so (Mhlambi 2012). Ngema became even more famous and much loved for his role in *Emzini Wezinsizwa* (The Young Men's Home), a television series about migrant men and life in eGoli (Johannesburg).

Was there any change in terms of the dramas' subjects and their treatment? I argue that there were huge shifts in the nature of what the dramas explored; that, in fact, because of the tenacity of their engagement with life as it was lived over so many decades, the writers of the dramas took their listeners deep into the trauma as well as into the excitement and glamour of '*uhlel' olusha*' and '*ikusasa lethu*'. Both writers and producers tried to catch the flavour of 'the new'. Yet, in some ways, as I will show, the *imidlalo yomoya* remained the same. It was, perhaps, a question of loyalties, but it was also an understanding that there was an aesthetic that needed to be respected and not tampered with too much, and a kind of moral economy with the idea of 'building' (*ukwakha*) at its core. As the title of this chapter suggests, it was a glorious decade in terms of radio drama, but it was often also full of pain.

In what follows I set out the immediate context of change for the medium of radio in the new era. Then, I address the nature of the serial plays that marked those years that were still known in the late 1990s as 'our tomorrow'. Thando Ngubane's remark in 2002 – she was then a Grade 10 student – that the 'stories' followed one another 'like life' gives us some clue to their continuing hold over listeners of different ages: the practice of listening had become a way of living one's life, in terms of moral decisions, crises, resolutions, emotions, and unexpected problems. The dramas, following each other in a long, colourful line, shaped subjectivities, showed people how to cope, and, most importantly, as many people stressed, allowed you to work out 'how to build your life [*uk' wakha impilo yakho*]'.[5] But what was different about *how* the dramas of the 'glorious decade' were 'like life'?

Media changes in the new era

Certainly, in the way in which the SABC functioned and was structured, there were drastic changes; these began in 1990 with the democratisation of the airwaves and the inauguration of a number of new commercial stations that were completely outside the sphere of public radio broadcasting. This shift to a broader democratic radio world was characteristic

[5] For instance, as recently as 29 October 2016, Faith Xaba of Mzimhlophe Hostel, Meadowlands, Soweto spoke of the dramas in those terms (personal communication).

of radio not only in South Africa but also in many other parts of the African continent, such as Ghana (Gunner, Ligaga, and Moyo 2011; Tettey 2011). Having been an important element in the attempt to control the airwaves nationally throughout the 1960s, 1970s, and 1980s (Coplan 1979: 149; 1985: 194), Radio Zulu was also part of the new dispensation in the media heralded by the Independent Broadcasting Authority (IBA) in March 1994. The station became known as Ukhozi FM, and was one of the nine African-language stations in the post-apartheid state and a media segment of the envisaged new egalitarian and multilingual society. The stations were part of the IBA's aim of:

> Ensuring that all official languages are heard throughout the country through the promotion of stations that reflect the realities of South Africa's integrated society, while guaranteeing the development and use of all its languages. (IBA 1995: 68)

The IBA also recognised the large audiences of the nine African-language radio stations, their substantial profits (in some cases), and their gross under-resourcing, and recommended that they be upgraded (IBA 1995: 68). Ukhozi FM – or, to use its full praise name, by which it had long been known affectionately by its listeners, *Ukhozi olumaphiko-phiko* (Black Eagle with Wide Wings) – had extensive regional coverage in South Africa. This reflected the use of isiZulu as first, second, or third language for many South Africans as well as its accessibility for Zimbabwean speakers of SiNdebele in South Africa who had close links with isiZulu through their historical connection to the nineteenth-century Zulu kingdom under Shaka (Omer-Cooper 1966). Many had maintained their ties to South Africa through decades of labour migration and shared traditions of music – both popular music and jazz as well as older forms.[6] Some came and stayed. The station's transmission covered the whole of Gauteng, all of KwaZulu-Natal, part of the Northern Province, and part of Mpumalanga Province and the Northern Cape; this wide coverage also implicitly recognised the status of isiZulu as a vehicular African language that potentially had great importance in the new era. A particularly significant part of its audience was the swathe of urban dwellers around Johannesburg and Pretoria.

The rural space

If radio was part of the drive to modernity, was it an urban modernity? Why did the rural space continue to be relevant? It must have done so in

[6] See Yvonne Vera's novel *Butterfly Burning* (1998) for a sense of the flows of music and popular culture between Bulawayo in the west of Zimbabwe and South Africa. For more on jazz and popular music, and the singer Dorothy Masuka, famous contemporary of Miriam Makeba and Dolly Rathebe, who hailed from Makhokhoba township in Bulawayo, see Gunner (1994).

part because forms of expressive culture, which themselves were modernising, were able to find a place on Ukhozi FM and mark a particular intimacy of culture, much as some forms of indigenous Aboriginal address used in indigenous media bound together distant and once shattered communities across northern Australia (Fisher 2016). Modes of expression were deeply multivalent, deeply heteroglossic (Bakhtin 1984), and held great symbolic value even in a fast-modernising cultural context. In some dramas, the space of the rural became a site for contemplating the possibilities of what the modern might be, and what it might have already become. More broadly, over the decades, radio drama had become a space in which transformations could be worked out, like popular culture itself (Stuart Hall, quoted in Samuel 1981: 232). The multiplicity of meanings in the mediated radio text could allow a piece to create its own world, yet still resonate with the historical moment – and, in fact, with contrasting historical moments. This point is important if we are to understand the impact of a drama first written decades earlier and then repeated.

Why should a play from the late 1960s, with its focus on the rural space, be re-broadcast in the 1990s? Was this simply nostalgia, and a radio station hedging its bets, as its listeners streamed to the urban areas to live, free of pass restrictions and far from the often underdeveloped and neglected rural areas? A case in point is the drama *Icala Lombango* (The Succession Dispute), written by Muntu Xulu and produced by Bhekinkosi Ntuli (the author and academic D. B. Z. Ntuli). In its first airing it ran for 15 episodes in July 1967, but it was programmed again in 1999, produced in the Durban studios under Masonto Buthelezi. In 1967, it had been broadcast at a time of increasing state vigilance in the Vorster era, with the security police in the shape of BOSS (Bureau of State Security) more oppressive than ever before, and the liberation movement about to move from 'stuttering sabotage [to] ... a more systematic better-funded military effort based outside the country' (Beinart 2005 [1994]: 213). In its first incarnation, was it simply a routine attempt by a state-controlled station to take listeners' minds off resistance? If so, why repeat it in a vastly different time?

In its 1967 incarnation, on one level *Icala Lombango* seemed to be the perfect play about apartheid and separate development. Its plot is simple, in a way stereotypical – but sometimes stereotypes are loved because they hold deep cultural resonance. The pages of *Ilanga* were sometimes filled with the dramas of key succession disputes, one being the famous case of who should succeed King Solomon in the early 1930s (Reyher 1999). Here, in *Icala Lombango*, Chief Shezi, from the Empangeni district of northern KwaZulu, has two wives; his favourite and senior wife has produced no sons whereas his second wife has. How should he resolve the dilemma of who should succeed him? We see him worrying about how to maintain his dignity when he is to attend a meeting of chiefs; he agonises over the succession question and decides to register, at the Bantu Affairs office, the name of his second wife's eldest son, a weak but pleasant youth. In the meantime his senior wife has fallen pregnant and the domestic air is thick with suspense. Will she produce a son? The second wife begins to plot in order to secure the succession for her son, and she and her family visit a famous local *inyanga* (doctor) for help. The *inyanga*, from Mozambique (some 300 kilometres due north of Empangeni), is called *Gwiny'inkomo* (Cow Swallower) and is enormously boastful, very expensive (as his name implies), and speaks a flamboyant and mangled Zulu. It turns out that he is also very inefficient, and, in the end, is proved to be a total fraud.

Finally, right prevails: the senior wife produces a baby boy and the young son is declared the rightful heir. The drama is tense, rich in its manipulation of character interaction and in its use of idiom and a kind of country Zulu that is very different from the fast, pared-down, slang-filled argot of the townships. It includes *izibongo* (praise poems) for poor Chief Shezi, clan songs, and many of the overt features of 'tradition' dear to the hearts of the architects of separate development. Yet, ultimately, it is a play about restitution and justice, and there is a sense of the ambiguity listeners could feel about the traditional features. They could see such elements as part of a multi-layered Zulu identity, one that interacted with many other kinds of experience and being, linking past and present in an extremely complex and radical way – perhaps a modern way.

Icala Lombango was also a play about language: the comical Zulu of the rich, immigrant fraudster, with its mixing of concords and awkward accent, emphasises the richness of the other characters' spoken language in the episodes. The drama illustrates superbly the issue of the 'multi-accentual' in language; on one level, the recitation of the *izibongo*, the use of clan praise names to address Chief Shezi, and the presence of other song forms such as the Shezi clan anthem (*ihubo*) feed straight into the ideology of the old rural homelands and apartheid-era Bantustans, where quiescent chiefs played the apartheid game. On another level, because of the richness of the language and the way in which it was able to summon up a pre-apartheid, resistant past, any project of domination was held in check. There is, for instance, a warning message of poison that comes in a dream to the childless wife. The warning comes from 'Your grandfather, Mbokod' obomvu'; he is thus named as a member of one of King Dinuzulu kaCetshwayo's regiments, thereby recalling the 1905 Bhambatha Rebellion and a moment of late resistance in which King Dinuzulu was tangentially involved. The dream thus conveys a far more fluid marking of power and territory, rather than simply being a narrative moment in the play.

The repeat of the serial in July 1999, 23 years after its first broadcast, saw it operating in a very different historical moment. It was repeated perhaps in an attempt to save on expenses in a strapped-for-cash but popular radio station with listening figures that, at the time, fluctuated between four and a half and five million per month.[7] Or it could be seen as a re-airing of a jewel of the 1960s, with a view to giving listeners a sense of the past and an understanding of the various layers of Zulu identity that were available and that could still offer a composite sense of identity within a wider, turbulent, and uncertain 'present' at the end of the millennium, questioning what a modern and modernising nation held for its citizens. Was it also a dramatic statement communicating the idea of a recuperative black modern, one in which 'tradition' could be selectively called up and edited in or out, as Karin Barber shows was done in Yoruba popular theatre (2000: 424)?

Youth, women, and rural emptiness

Perhaps it was the contradictory associations and possibilities of the rural space that were particularly held up for consideration in some of the dramas of the late 1990s. The rich and beautifully crafted Chief Shezi story, with its delight in language and a variety of speech registers, ran alongside the humorously dystopic drama *KwaNhliziyo Ngiyise* (Where My Heart Takes Me). This was followed shortly afterwards by Innocent Masinga's stark and powerful *Ngakubona Ngakubeletha* (Suffering! I Know You Well!). As companion pieces, the three gave a greater depth to the layered meanings of the rural and tradition and to modern life as seen at the cusp of the new millennium.

In particular, the two contemporary plays recast the space of the rural in a way that was completely in accordance with the rush to the city and urban spaces that marked the 1990s in South Africa. In the innovative *KwaNhliziyo*, the rural space is presented as a place of poverty of body and spirit and is linked to a clear message: follow your head and not your rash heart. The bold woman, Yoliswa, from Umlazi (in Greater Durban), who follows her heart and her attractive taxi-driver lover to his rural base – while he returns to the city – is appalled by what she finds in her new home in Nongoma in northern KwaZulu-Natal, known to be steeped in tradition and Zulu royal history. She is bored to death, complains bitterly about her diet of sweet potatoes and sweet black tea, and hates wearing '*ophaqa*' (the hard-wearing open sandals of the poor) and having to live without running water. When her long-suffering husband from Umlazi, about whom the drama has said very little, moves from the urban to the rural space to find her, he remarks in horror: 'You could plant sweet potatoes in the cracks in her feet!' So here, the rural space, far from being the home of the 'present past' or the 'storehouse of

[7] Interview with Khaba Mkhize, 1999.

the nation', holding the richly ambiguous symbols that might allow regeneration, is unambiguously unattractive. And the place of women within it – particularly those with ambition and energy – is shown as grey and unfulfilling. And yet, listeners might add, what came through in *Icala Lombango* was the forceful personalities of women, their eloquence and energy, even their ability to manage men in what appeared to be a hopelessly patriarchal gender hierarchy. Women, then, became markers of possible change and unquantifiable power in the difficult rural space.

The figure of youth, another unstable signifier in the post-apartheid period, also featured in the imaginings of the rural. In a drama that appeared very shortly after *KwaNhliziyo* (Where My Heart Takes Me), the rural areas have become a place of nightmare, and the focus is on disempowered youth. In the stark and powerful *Ngakubona Ngakubeletha* (Suffering! I Know You Well!), with a gnomic popular proverb for its title, the rural space is filled with terror, danger, and anarchy. The visceral link between the struggling body and the idea of suffering is sewn together in the Zulu version of the proverb, which translates literally as '[Suffering,] I have seen you, I have carried you on my back'. Nothing could be more different from the ambient richness of Chief Shezi from Empangeni and his worries over a son to succeed him, the comic and greedy doctor (*inyanga*), and the determined young wife – characters and themes that were all part of the narrative pull and pleasure of *Icala Lombango*, the 1967 play that was repeated in September 1999.

In *Ngakubona Ngakubeletha*, the rural setting is never specific; rather, it suggests a possibility to be fleshed out with the listeners' own experiences and references. It is a desert of social anomie. There is no chiefly rule, no homestead life of any kind, only fear. There is, in fact, no rule or sense of order at all, apart from that of the power of an older man, and the gun. The rich comfort that listeners found in the 'thick' and idiomatic language playing over the events in *Icala Lombango* is absent. Instead, the listener keys in to the asocial, the aural motif of galloping hooves counterpointed with forest sounds. A half-mad horseman gallops across empty terrain searching for the youth he wishes to kill; the young man in question, who is innocent of any crime, crouches in limbo in the forest, an outcast waiting to avenge his family's destruction. He talks to himself, hidden from and outside any social order or family comfort. The ending of this drama tied up the loose ends in a rather perfunctory way, and the young man eventually found security and a family of a kind with his sister in an unnamed township. Yet what resonated most powerfully was the play's constant evoking of a destabilised family unit, the barrenness of the rural as an emotional or physical anchor, with no hint of harmony or richness.

Ngakubona thus possibly set out for its radio drama listeners the symbolic landscape of the civil war violence of KwaZulu-Natal in the mid-1980s and early 1990s. But, if read allegorically, or 'multi-accentually', it spoke very precisely to the situation of youth at the end of the 1990s. It captured the isolation and structural repression, particularly of young men, who, by the end of the decade, were affected most by the unchecked and largely unacknowledged AIDS pandemic, had poor or no job prospects, and had little hope of marriage according to established social norms (Hunter 2010). It was a generation that was disinherited and in limbo, like the fugitive youth in the quite apocalyptic *Ngakubona*. Could this dystopian drama also have been an oblique comment on what the black modern could become: a modernity devoid of any real signs of improved material life or opportunities, a dark shadow of any real, meaningful modernity? This is a point James Ferguson explores trenchantly in *Global Shadows* (2006).

Aired only two months after *Icala Lombango* (The Succession Dispute), the dramatic airspace of Innocent Masinga's script for *Ngakubona Ngakubeletha* (Suffering! I Know You Well!) and its affective sonic components of terror and despair offered its audiences an aurally compelling account of a bleak and disinherited present, particularly for young men. It was a dystopian vision at odds with the then still current optimistic phrase of '*Ikusasa lethu*' (Our tomorrow). It sat, suggestively and powerfully, in the radio repertoire of the station's still loyal audiences, just after the replay of 1967's *Icala Lombango*, which gave its listeners an entry into a more stable world, a past recalled with its own rich and ambivalent resonances and its own set of ambiguities – around gender and women's rights, for instance. Listeners had a chance to compare, to ponder, and to debate the value – and needs – of the rural setting in the bewildering years of the late 1990s, and to enter into discussions about what precisely constituted a good man, a good husband, a good marriage, and a good life in this new era. I return to this point later in the chapter. First, though, the manner in which the dramas focused on young men needs further exploration.

In addition to a focus on the barren rural space and the past and present cycles of violence in people's lives, one of the themes that surfaced constantly in the dramas of the late 1990s was that of a lost son, or lover, or brother. In a number of dramas this was done in a very graphic way, with a great show of the agony, sorrow, grief, and sense of loss experienced by the family, lover, or wife of the lost youth. The recurrence of the lost son/brother/lover story and the fact that audiences seemed to need to have such stories suggest to me that, in order to deal with this subject in the public/private narrative/dramatic form of the radio drama was very cathartic on some deep level. The grieving – because that is what it was – could be both deeply personal and at the same time communal through shared listening over the airwaves. The terrible violence in KwaZulu-Natal from the late 1980s and into the 1990s, and the fact that in many cases it was young men who died, may in some way have been addressed through such dramas. The precise details of the storyline may have been immaterial; what mattered most to listeners was following the passage of emotion through love, loss, and grief. The difference between the dramas and real life was that in the former, in most cases, the son was returned and there was a renewal of harmony at the end of the story. This lack of 'truthfulness' compared with the actual history of loss was, I argue, in keeping with the generally upbeat and optimistic nature of many, though not all, of the dramas. They did not seem to lose any of their power to captivate and hold their audiences as a result. The theme of the lost son remained dominant throughout the 1990s; this may have been because of the loss of youth through the AIDS epidemic in KwaZulu-Natal – and more widely in the region covered by Ukhozi FM. But the theme might also feed into other significant storylines as well, storylines that have a link to oral narrative. In a 1997 radio drama by Cecilia Mkhwanazi, *Impic' Abadala* (A Tricky Situation), the story narrative was built around two lost sons – twins called Vela and Velaphi who are lost when young, when there is fighting between factions in a rural area. As their mother crosses a flooded river, one of her sons is left behind and supposedly drowns; the other runs away when he hears of the violence. In fact, both twins survive, and years later one is an accountant and the other a hobo living rough. The hobo becomes a criminal, but the accountant is arrested in his place and jailed. Eventually the mistake is discovered and the two brothers confront each other – mirror images but caught up in different social trajectories – which was how the drama ended. Such an open ending invited listeners to dwell on the way in which fortune could lift you up or fling you down; that sometimes success was in your hands and sometimes you were helpless against fate. The drama could also have been seen as a comment on the times, their uncertainty, and the sharp new contrasts of social status and economic well-being that were beginning to emerge as neoliberal policies struck home (James 2012; Hull and James 2012).

The regime of money

The pleasure and expectation of consuming and consumer culture were noted as excitements of the new era, particularly for formerly disenfranchised black subjects (Posel 2010), and this was reflected in the radio dramas of the period in a number of ways, some of them unexpected. Greed for money had been a powerful element in early dramas such as *Umcebo Owalahlekayo* (The Lost Treasure) (1968), in which K. E. Masinga played an oppressive father figure and that featured a fugitive son called Mandla (Strength), innocent and wrongly accused of stealing the buried money of his adoptive father. This was a favourite, often repeated. But in a few of the dramas of the new era, greed was linked to the new times and the urgent need for money in novel ways. The September 1999 drama *uFakazi* (The Witness), by M. C. Banda, ran alongside *Ngakubona*. However, its focus was very different. An urban drama, it set its sights on gangsters, good men, and retribution. It thus kept to the old pattern of 'working through' a complex moral problem that had kept the serial dramas alive for decades. It also introduced another well-established device, but gave it a modern twist: the spectral world of the deceased penetrated the fast, slick life of the wealthy gangsters who had callously murdered their business partner, a good man who had put up most of the money for a new commercial venture. Here, the spectral and otherworldly, which represented the world of truth and morality, came in the form of a spirit cat, 'the witness' (*uFakazi*) that gave the drama its title. The uncanny meows of the spirit cat pierced the drama's soundscape in a way that was quite unnerving for both the gangsters and listeners, reminding them remorselessly of the good man they had killed. Through the spirit cat and its sonic interventions justice strikes home; in the fast new age of 'our tomorrow', with its pared-down street argot and gangster talk, greed – so the drama concluded – is a moral cul-de-sac, a dead end.

Money, integrity, and relationships also dominated a fairly short serial drama (of only 27 episodes) in the early months of the new millennium. But here a new component was powerfully present: 'the community'. *Isiboshwa* (The Prisoner) was not about the closed world of crooked gangsters or businessmen, but rather it asked: 'What happens to the community if those who are meant to be moral and uphold public integrity go rotten?' There were, though, both links and differences with *uFakazi* (The Witness). The lure of fast money and a gangster lifestyle was more apparent in the second drama. The young gangsters – Sbu, Lindo, and Mbuso – who make their money from car hijacking and car thefts are shown as living very well and in grand style. They speak the new, pared-down Zulu argot of the streets, live fast, take risks, and represent a strain of quite powerful youth masculinity circling outside the morals of the community yet feeding off it. Set against this potent dynamic is a solid, older policeman, Makhaza, who is determined to bring the township of Umlazi on the southern edges of Durban back to what it had been in its early days. It is Matha, the young wife of the weak and vacillating 'rotten policeman', Mazwi, who speaks the phrase made famous by the Lucky Dube reggae song of a few years earlier – 'There's no truth in the world!'

The added twist in this serial was that it showed with great care and in the 'slow time' made possible by the long, steady progression of the dramas that Matha herself was partly to blame for her husband's loss of integrity. The drama hints that her passion for luxury goods and a better lifestyle – linked to studying and moving up the class hierarchy – plays a part in her husband's decision to work with the wealthy predators. Whereas, by the drama's end, the doomed Mazwi realises that he has lost all integrity, his wife never achieves the same self-knowledge: that her restless desire for a better life may have played a part in her husband's decision to turn crooked. Matha can only say, weeping, as he lies dying in hospital from a suicidal gunshot, 'What was the problem! I gave him all the love of a wife!'

Possibly *Isiboshwa* judges the young wife too harshly. Yet of great importance was the manner in which it foresaw, and laid out very clearly, the possibility of a complete derailment of the 'our tomorrow project' of the new era, if public integrity were to be compromised by greedy or weak men. The township of Umlazi, the community – *umphakathi* – of which it consisted, could stand metonymically for the state. *Isiboshwa* had the typical sound features that were characteristic of all radio dramas – the signature tune, *ibika,* and then a choice of fast music bridges that heightened affect, depending on the scene to come. Thus the dramas were always 'sound events'. However, *Isiboshwa* may have had a particularly potent combination of producer and sound engineer. Muvo Maphalala produced it and Dumisani Nkosi was responsible for the bridges. The lush, fast tango that was its signature tune, the bridges of brief, plangent guitar chords, and the sorrowful harmonica with orchestral backing were all part of the total 'sound event' of the drama. Only at one point does music have a part as an 'actor' in the script: in the court scene very near the end, when the crowd, the important *umphakathi*, waits to hear the judgement, reggae music is playing. It is present but unobtrusive, representing perhaps a link between reggae as a form of popular music and its history of rights and people's justice.

Some dramas did not become well known and their names were never on the lips of listeners, but that did not mean they had no impact. In the way in which they allowed listeners to engage with the twists and turns of shifting social values and norms that the later 1990s in particular opened up, they can be seen as simply putting on record, in the serial drama form, the conflicted moral possibilities underpinning the choices of everyday life. One such drama caught well the new excitement over money and consumption and the sharp class differences in black society, which, by the end of the 1990s, were clear for all to see. It may never have captured listeners' affections because it was too full of incident. Certainly, unlike some other dramas, it never became the focus of listeners' conversations, and repeats were not constantly requested. It was possibly too exhausting with its constant demands on its audience to react with horror, anger, sympathy, rage, anxiety, and so on. All the worries, sorrows, and fascinations of the era were crammed into it. The narrative traced a botched schoolgirl abortion that led to death; a teenage pregnancy and then a birth, with the new infant seeming to have the features of each of the three possible fathers, all married men; a love affair across classes in which the indulged but strong-minded daughter of an affluent taxi driver falls in love with a youth with no prospects but a loyal heart; a wise father who is a *sangoma* (diviner), with a weak son who is a minister of the church. Moreover, the drama involved faith and fashion; a feckless, scheming mother; a distant but socially ambitious professional mother; guns, thugs, and fancy hotels; tomato juice suspected of being a love potion (*isiphuzo*); lost bank cards; and AIDS lurking in the background – all these and more featured in a drama that ran from the end of 1999 into the first month of the new millennium. It was called *Kungenxa Yakho* (It's Your Fault) and was written by Lovemore Wiseman Ndlovu. In its determination to highlight a total loss of any fixed moral centre or sense of personal responsibility, the script has a number of the characters say as their lives fall apart, '*Kungenxa yakho* [It's your fault]!' What needs to be remembered is Thando Ngubane's comment at the start of this chapter: 'One story follows another, like life.'

Family

Certainly, *Kungenxa Yakho* (It's Your Fault) placed the family, or different kinds of possible families, at the centre of its plot and its plentiful, excessive emotionality and melodrama. In this it was typical of the genre. The family was without doubt the symbolic site that the playwrights of the serial dramas, watched over and 'managed' by their head of drama, continued to mine, in their determination to set out, dissect, and explore the terrain of 'the new'. Sometimes, as in the case of the overloaded *Kungenxa Yakho*, the writers used particular scenarios to weigh up different knowledge systems by comparing, for instance, the wise diviner with his superficial pastor son.

Other dramas from the same era handled the site of the family very differently. One was a drama that would become a great favourite with listeners, who constantly asked for it to be re-broadcast.[8] *Ebuhleni Nasebubini* (For Better and for Worse) focused on a single marriage. Written by the prolific and versatile Dumisani Nkosi, an author as well as a master of radio drama 'sound', its first part ran for 34 episodes. The producer was the veteran of radio Joshua Mlaba, and the cast was filled with famous names such as Eric Ngcobo, Lawrentia Madlala, Shadrack Ngema, Vicky Masuku, and Dalton Ngwenya. Listeners pleaded for a second part. Reluctantly, Masonto Buthelezi agreed, and after a few months Part 2 aired. When Wiseman Masango and I questioned

[8] It played again as recently as September 2016.

Nkosi, the author, about the drama,[9] he explained that he had wanted to address the question of the importance of the relationship between spouses, whether or not there were children from the marriage. It was a story full of violence, tenderness, love, and betrayal – with atmospheric linking music, which was typical of all Nkosi's dramas, and, indeed, a mark of all the radio dramas and a part of their 'grammar'. Two people fall in love and decide to marry, but when the wife does not fall pregnant the husband falls in love with another woman, marries her, and then discovers that the child he thought was his is the Indian foreman's. He kills his second wife, goes to jail, and wants his first wife back. That was the point at which Part 1 ended. The listening audience, the 'knowledgeable community' hooked on the genre, felt that the story was incomplete and demanded that it continue; and so, eventually, it did. After a great deal of serious melodrama and tension, good triumphed.

It would, I argue, be short-sighted to see the audience's desire for more, and for a kind of resolution, as escapist or as indulging in fantasy. Instead, the story being played out allowed a kind of public–private meditation around the issues of love and fidelity, issues that story raised as questions for the 'new era' (*uhlel' olusha*) while also allowing a look back to and a reassessment of 'tradition' as the frame of marriage. In that sense, it cut across the intersecting circles of marriage laws in the country: marriage according to customary law, or civil and religious law. It had a central focus, clear of clutter, and addressed a question as important to listeners as those dealing with integrity and corruption in the public domain. It was also, clearly, intensely enjoyable. The rich language of the play, the music cues, and the individual encounters, such as the visits to the *inyanga* (doctor), allowed listeners to engage on different levels with different aspects of modern life as they were living it. And they were living that life in a very different 'tomorrow' from the one seen through the utopian lens of the expected 'new era'.

Thus, the clutch of late 1990s dramas discussed above, with their variety and their voracious engagement with life on the ground, held no single solution. Instead, they asked – perhaps even forced – listeners to grasp their narratives and engage fully with them as people who were involved in making life happen, and who had a role to play and an interest in whatever was offered in the new dispensation. Most of all, through this art form, listeners knew that they were not passive absorbers but active participants. It was a heavy responsibility for any genre, but it was one that the plays of the air, *imidlalo yomoya*, carried through the latter part of the old century and into the new millennium.

[9] Interview with Bheki Msani and Dumisani Nkosi, 1998.

8 Finding a centre

> Don't be like a solitary cow, grazing on its own and no one knows to whom it belongs. Pastor S'thembiso Zondo on the *Vuka Mzansi* Breakfast Show motivational slot on Ukhozi FM, 8 June 2014[1]

Becoming part of people's new public selves

Pastor Zondo's early morning exhortation to listeners of Ukhozi FM, as they rushed to work or went about their business, was delivered in his recognisable fast, upbeat verbal style. It can be seen as summing up a wider message about belonging in an age of bewildering changes and conflicting demands. Pastor Zondo's swift sound clip was intended to strike each and every listener with equal intensity. Each would interpret it in his or her own way. How, and to what, could one belong? Because Zondo was saying – not only to his Durban listeners but to everyone who was tuned in to the station at that prime time, in all parts of the country – 'To belong is a must.'

In this chapter, I ask how a radio station that had taken its listeners through so much in the apartheid years could continue to build a centre, or a series of centres, in a vastly different historical moment. I suggest that, in the new democratic era, the intimacies that previously had largely been contained within the dramas became part of people's new public selves. They were voiced by listeners on phone-ins to Ukhozi FM, and in particular to the *Vuka Mzansi* Breakfast Show, which I discuss in this chapter. Also, the endlessly displayed selves in the domestic dramas in which the site of the family was constantly tested and explored could now speak as individuals and tell their own stories on air. In a sense, then, to draw from Kunreuther (2014), they became 'voicing subjects' within a wider democratic space made possible by radio. This does not mean that the space of the domestic was lost within radio dramas themselves. The shifting site of marriage as a place of tested loyalties and conflicted

[1] Pastor Zondo, a well-known and esteemed Durban preacher and personality, was a frequent speaker on the *Vuka Mzansi* Breakfast Show in 2012 and 2013 and again for a time in 2014.

ambitions continued to be explored in the final drama discussed in this study. The intimacy, affection, and loyalty that the station had forged, against the odds, as Radio Bantu, Radio Zulu, and then, from 1994, as Ukhozi FM were not states of mind that were set in stone. The single and multiple communities created, the listener identities that came into being through radio soundings, were not static, as this widely researched topic has shown (Crissell 1986; Douglas 1999; Hendy 2000; Gunner, Ligaga, and Moyo 2011). Cultures of listening, like particular radio programmes or genres, are movable feasts. How did Ukhozi FM re-imagine the multiple transitions, the tensions, the pleasures, and the suffering of its community as both the station and its listeners continued on their journey in the second decade of the new millennium? How did it, perhaps, shape listener identities in a way that helped them *live* in the era of sudden change rather than simply exist? This is the central question of this chapter. I examine first a powerful serial drama that probed the meaning of marriage, modernity, and tradition; for so many years, the dramas had provided listeners with ways of imagining the self, the family, and the social through a resistant cultural form with a distinct aesthetic. I then discuss the presence and 'pull' of the important *Vuka Mzansi* Breakfast Show as it helped listeners create a new imaginary, a centre that might hold. Could this particular programme, as a part of the station as a whole, 'perform' and 'voice' democracy, as Nick Coldry (2013) has asked of the present role of the media? And, if so, using what means? Could it be through the 'voicing subjects' of the now not so new democratic era? Or do we need to cast our conceptual net more widely?

A haunting presence

The writer and radio man Morris V. Bhengu (also author of the unsettling *umdlao womoya*, *Yiz' Uvalo* (In Spite of Fear), discussed in Chapter 6), wrote what was to be his last radio serial in 2000–1.[2] Three years later, *Bakikiza Buthule* (Wedding Dances in the Silence) played for listeners. Like other powerful serial dramas of Radio Zulu/Ukhozi FM, it had repeat airings, its most recent being in August 2016. *Bakikiza Buthule*, with its unsettling title, did not in any obvious way stick to the criteria that Masonto Buthelezi had spelled out so emphatically for Wiseman Masango and myself 17 years earlier, as he sat at his Durban desk as Head of Zulu Drama: 'No Part 2, no divorces, no shooting. But

[2] Bhengu had won the prestigious M-Net literary award in the Zulu fiction section in 1998 for his novel *Itshwele Lempangele* (The Guinea Fowl Chick). The serial play discussed here was the last of 15 written by him for the station's series of radio dramas; one of the others was *Yiz' Uvalo* (see Chapter 6).

I don't like flat dramas, make no mistake!'[3] Rather, Morris Bhengu's serial made its own rules, while acknowledging the conventions of the genre and in key ways being guided by them. Was there a way in which it 'held the centre' for the tradition from which it grew, but at the same time set up new, difficult questions that threatened to destabilise it? Its focus on marriage and the family went to the heart of the very being of the *amastori/izitori* serial dramas. Yet, in capturing the uncertainties and angst of an era, it presented dilemmas that could not be resolved within the usual frame of the dramas. Perhaps in this way it was close to interpreting the spirit of the times, which was agonistic and despairing yet aspirational. I argue in this chapter that one way of reading the station as a whole, 20 years into the 'new era' is to see it as celebrating with a kind of inspired sensibility the strands of creativity and energy, together with the dangers, that continued to drive listeners and broadcasters forward, however demanding the present.

A theme running through the chapter is the role of language, and the Zulu language itself as a flexible and versatile 'character' (Ndlovu 2011) in the broader drama of the era, at a time in which the national, the regional, and the global all stamped their mark on the mediations of the station. The terrible strain, the skewed newscasts, and the silences of the apartheid era – silences that newsreaders such as Thokozani Nene had tried to challenge and others had suffered through (see Introduction; Chapter 5) – were gone. How did the radio soundings mediate and inform, working through the Zulu language and the voices of the presenters, who were often glamorous role models for the aspirant listeners of the new millennium? How did they voice the thoroughly changed politics of the personal, the social, the sexual, of what came after the demise of apartheid? How did they maintain a centre that had a place for the increasing number of the marginalised and the destitute, alongside the working classes, the aspirant, and the middle classes, and those who made a living in a variety of ways, both formal and informal, but who all lived their lives, at some level, through the medium of the language of Ukhozi FM, isiZulu? Had the language become an important player in national voicings because of the long and sustained role it had carved out for itself, from the apartheid era to the present?

A proper marriage

In setting *Bakikiza Buthule* (Wedding Dances in the Silence, or just *Bakikiza*) in the freehold suburb of Clermont, which was marked clearly as part of the modern coastal city of Durban, with its beaches and its traffic jams as well as its bars and its domestic and office spaces, Morris

[3] Interview with Patrick Masonto Buthelezi, 1999.

Bhengu established the spatial and emotional topography of a modern family. In this, he was being true to one strand of the long-established aesthetic of his chosen genre, the drama of the voice. It was a statement, too, about the long presence of an 'integrated [South African] urban modernity as a fact of life' (Kruger 2013: 44), and an African urbanity with deep roots in the city (La Hausse de Lalouvière 1989; Erlmann 1991). The drama also revealed, quite unambiguously, the presence of custom within marriage in this modern African city and examined it with a critical but sympathetic eye. By the end, listeners had been forced to ask themselves what the limits of custom might be, and also to consider the dangers of an unmoored life – where, like Pastor Zondo's 'solitary cow' – one belonged to no one and nowhere. Was the drama casting a look back at the community created by the station in its long history? Could it sustain this in the fragmented cacophony of voices of the new times?

The marriage of Dolly Dlamini and Maviyo Khumalo, the central characters in this drama of a 'proper marriage', was based within the norms and rites of *custom*, with church rites present but in a more distant, if still important, way. In the final moments of the 50-episode serial, it is the space of the church that holds the characters in death, even as it formalised their marriage through the wedding ceremony, and Dolly's mother has the last word (Episode 50). The serial drama tells the story of the tempestuous and short married life of Dolly – MaDlamini – and Maviyo Khumalo, as they attempt to navigate questions of authority in their relationship, address the need for money to keep the home afloat and pay a bond, and deal with the world outside, and the world within. The latter is signified by the space of the home, and the outside world, variously, as the bar or tavern, the work space and its people. Forming an intermediate space, sometimes framing and attempting to control both the world outside and the world within, is the extended family in the form of Maviyo's older brother, Bhut' Simon, and his wife, MaMthembu, and Dolly's mother, MaSithole, although the husband's line carries more authority within the family.

Importantly, the influence of the extended family is not shown only through its 'present' representative, Bhut' Simon, but also through the weight of custom: the ancestors of the two families were joined in the marriage, and the *lobolo* cattle were paid. In that sense, then, it is a 'proper marriage'. Constantly probed, however, within the narrative arc of the story are the rights of husband and wife in this marriage of the 'new era', which includes the marked presence of new laws enforcing women's rights if there is abuse in the home. The ending of *Bakikiza* is a tragic one, with no systematic working towards a resolution, unlike most other dramas. Yet the shadow of such a resolution hovers over the drama as a failed possibility. The rage of Brian, Dolly's lover, ends in the fatal shooting of the estranged husband and wife as they speak together, in their old home. They die at the very moment when they were possibly moving towards a reconciliation and a new understanding of what a 'proper marriage' might be.

A conversation with Dennis Mseleku, sound archivist at the Durban SABC, about the drama and its 2016 repeat a few months earlier alerted me to the way in which *Bakikiza* could draw in a new generation of listeners, as well as engaging a more seasoned audience, part of its established listening community. On one occasion, in a taxi from the centre of Durban to the suburb of Umlazi, he overheard two older listeners, both women in their fifties, praising the very high standard of the actors in the drama, all of whom were well known to them. On visits home to rural Umzinto, south of Durban, over the two-month period during which the drama was broadcast in 2016, Mseleku found a similar preoccupation with *Bakikiza*, but in this case among first-generation listeners who were intrigued by the problems that Dolly and her husband faced, the drama's intensity, and the way in which it captured what seemed to be the insoluble challenges that they themselves might face.[4] Wiseman Masango and I had listened to it on its first airing in 2001, but its true force only struck me when I heard it a decade or so later in one of its replays. Its aptness – held partly in its eloquence and the depth of its writing – in setting out for listeners the heightened strain on marriage, the problem of money, and how to work out a life in the second decade of the millennium gave it an added power. Because of the range of the voiced drama, Bhengu's script could reach an audience of perhaps half a million, a feat impossible for a novel or even a popular stage drama. This fact has to be factored in to the impact of voice and radio.

The strong characters of both wife and husband are evident from the very beginning of the story. The sweetness and homeliness of Dolly, who sings as she cooks her husband's meal, a little later than he expects, and her diligence as a seamstress are in sharp contrast to her steely resolve to persevere with her demand for the right to seek work, even when she is opposed by an abusive and often drunk husband (Episode 2):[5]

MAVIYO: Why're you standing doing the cooking now? What've you been doing all afternoon? [*Kwangathi yima upheka nje? 'Bewenzani yonke imini wena?*]

DOLLY: I've been sewing, Maviyo. I had a lot of sewing to get through; I've had to get ready for pensioners' day and the pinafores for the old ladies. [*Bengithunga Maviyo. Muningi kabi umsebenzi wokuthunga ongihlalele, ngizama ukuhlinzekela usuku lwempesheni yezalukazi.*]

Dolly then tackles her husband on the question of money, the upkeep of the home, which has fallen on her shoulders, and her desire to go out to

4 Personal communication with Dennis Mseleku, 2 December 2016.

5 Here, as in the other dramas, the scripts I refer to were constructed from transcriptions by Wiseman Masango, which were typed up. In this case, however, I listened to the drama and also heard it again in one of its recent airings.

work, to sustain the home – and, it is implied, to nurture him (the verb '*ondla*' stresses this). As Maviyo ducks the question and attempts to leave the house for his favourite tavern at Sis' B's, Dolly blocks the door and urges him to answer her:

DOLLY: Hey! Don't run away from the truth, Maviyo ... Say I can work; just come out with it! ... No woman could keep her sanity living the way I'm having to ... Come on then, don't be afraid. If you want to hit me, hit me! [*Heyi*! *Mus' ukubaleka ngeqiniso Maviyo* ... *Ngivumele ngiyosebenza wena; phumela obala*! ... *Akekho umfazi ophilile lapha ekhanda ongavum' ukuphila lempilo engiphilayo*... *Uze ungazisoli. Shaya*! *Shaya*!]

At this critical moment, the calming world of the extended family breaks in, with the entrance of the steady older brother, Bhut' Simon, with his standard formal address – '*Sikhulekile ekhaya kwaMntungwa* [We greet the home of those of Mntungwa]!'[6] Here, the formal salutation with its respect for 'home' (*ekhaya*) and an invocation of Mntungwa – the Khumalo family praise name – cools down the ragged emotionality that precedes it. This is suddenly not about just two individuals but is a statement of their existence in a wider temporal and situational frame. Also, the focus turns to a key and multivalent concept that is crucial in marriage, and in Zulu cultural norms: respect, *inhlonipho*. While Maviyo complains of his wife's lack of respect, in the course of the dispute his older brother reminds him that respect in a marriage has to be mutual:

MAVIYO: The *respect* due to me, as a husband, from my wife is utterly and completely gone! [*Uphelelwe nje nya inhlonipho yami njengendoda yakhe*!]
BHUT' SIMON: Respect is a two-way deal, Maviyo, My brother, do you hear what I'm saying? [*Kuhlonishwana kabili Maviyo, Mfowethu. Uyangizwa*?]

As Bhut' Simon leaves, his mediation may have failed. Nevertheless, the weight of his words and their *systemic* authority carry through to the episodes that follow, as Maviyo destroys the basis of marital respect between two individuals through his physical assaults on Dolly. He refuses to countenance her working outside the home and attempts to pound her into submission. The radio sound text conveys graphically the violation of Dolly's body and her dignity. Yet it does so by showing it at one remove, as the daughter, Dolly, recounts to her mother, MaSithole, what took place. The telling thus has the force of testimony – like that of a victim to a public authority, the police, or a court of law, where, in the new dispensation, such abuse could be reported and punished. It also, though, has the affective power of the intimacy between mother and daughter. Lindiwe Ntuli, who plays Dolly, is recognised as one of the

[6] 'Mntungwa' is the main term of praise and respect, *isithakazelo*, for anyone with the surname Khumalo, and carries with it a sense of the past in the present, a shared history of lineage, and therefore a shared present.

great actresses of the Radio Zulu/Ukhozi FM drama tradition. Surely, she was one of the actors whom the speakers in the taxi to Umlazi had in mind as they commented on *Bakikiza* and on the power of its acting, so typical, they said, of the earlier *izitori* (dramas).

Yet nothing is straightforward in this twisting narrative of marriage and family, which acts in some ways as an ironic intertext with other earlier dramas, such as the idyll of love and courtship *Ubongilinda Mzikayifani* (You Must Wait for Me, Mzikayifani), which first played in the 1970s (Chapter 5), or the bucolic and much-loved narrative of *Icala Lombango* (The Succession Dispute) (Chapter 7). As the police search for Maviyo (in Episode 5) to press charges of domestic violence, there is a plan for Dolly – MaDlamini – to drop the charge if her husband agrees that she can go to work. Only after this plan is discussed are listeners allowed to experience the pain of the assault itself, as described by a daughter to her mother. The account, and the listeners' shock at knowing that it is never likely to reach the police, sit tight together in the text. Affect – and the intimacy of mother and daughter – intensifies the sense of the broken and bruised body of Dolly. At first, the 'slighter' physical damage emerges. Dolly, unable to move, can't drink the coffee her mother, MaSithole, offers her, and she asks for ice:

DOLLY: Please can you come closer. [*Awungisondeze.*]
MASITHOLE: What's it, *sisi,* what d'you want, my girl? [*Yini sisi, ufunani ntombi?*]
DOLLY: Ice, Mama, that … [*U-ayisi Mama lowo …*] Please can you look here at my lip, Mama, to see if the swelling has gone down? [*Awungibheke lana odebeni khona kuzokwehla ukuvuvukala Mama?*]

Dolly then describes how Maviyo hit her face first with the palm of his hand, then, with greater force, with his fist. As her mother becomes increasingly agitated, she describes how when she was on the ground he kept saying, 'You'll listen to me! You'll listen to me! [*Ubelokhu ethi, 'Uyangizwa, uyangizwa!*]' The assault is not over. Dolly continues: as she slid under the bed to escape he mauled her with his hands:

MASITHOLE: My child nearly died! Oh, oh Jehovah! [*Cishe yafa ingane yami! Jehova, hha … wu!*]
DOLLY: I don't know, I really don't know why he had to still go on and bite me. [*Angazi, angazi-ke ubesangilumelani.*]
MASITHOLE: Oh to think of it – the imbecile – that a man could bite his wife! [*Awubheke nje, isilima sendoda, ukhona umuntu oluma umfazi wakhe kanje!*]
DOLLY (CRYING): He bit into me with his teeth here, on my back. [*Wangidla ngamazinyo lapha emhlane.*]

Despite MaSithole's anger and sorrow at her daughter's wounding and her comforting words to Dolly that the case is safe in the hands of the no-nonsense policeman Hlophe, listeners know that it is not safe at all, that a deal may still be made and the case withdrawn. If that happens,

Dolly will gain her wish to work but justice will not win the day. Episode 5 ends on a note of daughterly vulnerability, motherly love, and listener suspense, and, for some, outrage:

DOLLY: Please put the light out, Mama, it's hurting my eyes. [*Ngicela ucishe isibani Mama, siyangixopha lana emehlweni.*]
MASITHOLE: Sorry Ma, sorry my baby, sorry *muntuza kaMa* [mother's special baby].

The explicit and visceral details of Dolly's assault at her husband's hands – and how the event plays out within the larger dramatic narrative – may have no close equivalent in any of the Zulu radio dramas I discuss in earlier chapters, yet what is striking is Bhengu's fidelity to the genre as he simultaneously pushes to take further his meditations on the family in times of acute change and on social stress in a particular era – and also to interrogate tradition and its value. Violence in the family and the question of disputed authority, as Winnie Mahlangu's early 1960s serial drama *Buzani kuBaba* (Ask Father) had shown, were no strangers to the radio genre.[7] Similarly, Abigail Zondi's 1970s serial *Abangane Ababi* (Bad Friends) dealt with the question of bold women who wished to push beyond the confines of accepted conduct within the broad ambit of tradition. Yet here the spatial and emotional 'transgressions' of the central character are quite lightly, but very skilfully, drawn, and at the end she realises the dangers of unknown paths and resolves to marry her remarkably docile fiancé (Chapter 5). The importance of the voices of older women as strong advisers, companionable gossips, and comforters also has a significant place in Zondi's drama. Similarly highlighting the intergenerational bonds between women, Alexius Buthelezi, working with the related genre of the music drama in the prize-winning *uNokhwezi*, was able to weave in as a sub-motif the closeness of mother and daughter, as Nokhwezi's mother, prompted by a messenger from the past, gives her daughter the necklace for protection in times of great danger (Chapter 2).

Interiority, and the key soliloquy that allows the listener to feel shock or sympathy with a particular character, was also characteristic of the techniques of empathy and illumination that writers – and their producers – made use of, and still do. Few, however, pushed the development of character to the level to which Bhengu took it in *Bakikiza*, as he set questions for listeners about modern marriage in the present but also looked back, folding in both custom and 'the now'. Few, too, probed the meaning of masculinity within the context of the past or the present as forensically as Bhengu did in *Bakikiza*. The pushing of boundaries within the whole tradition of the dramas/*izitori* was, I argue, a way of

[7] The isiXhosa version of the same drama, *Buzani kuBawo*, is discussed by Khaya Gqibitole in his PhD thesis 'Creativity or control?' (2012).

searching for a new centre at a time of social stress, even trauma, and of achieving it through a range of vocal registers.

'When a man cries/It must be something very terrible'[8]

The role of Maviyo, the reprobate and conservative scallywag of a husband, who, in spite of everything, is not a bad man and becomes a tragic figure by the end of the drama, was played by an actor revered by listeners, and possibly the most popular of all the rich cache of voices on Radio Zulu/Ukhozi FM. This was the late Shadrack Ngema. The strength of the acting illuminated to maximum effect the complexities of the character. Maviyo's weaknesses, coupled with his ability to realise his mistakes, become central to his appeal to listeners. Bhengu manages the seemingly impossible and wins back listener sympathy for Maviyo, through a series of plot moves that show the strong-minded but naîve Dolly unable to negotiate some aspects of the world of work, in particular the predatory advances of her boss, Brian Mkhize, played by Eric Ngcobo.

The strength of Maviyo's love for his wife is conveyed through his journey to self-knowledge and his realisation that he has lost everything – his work, his wife, his self-respect – as he acknowledges to himself that he has become a habitual drunkard. He, too, is vulnerable. He passes out in public when he is told that Dolly has chosen another man (Episode 30). This news is filtered to the listener through the words of Momsy, Maviyo's drinking partner, as she tells him what has happened:

MOMSY: Ah! Darling, you've come to! [*Hhawu, usuphapheme mntakwethu!*]
MAVIYO: Yes, Momsy, I'm awake. [*Ya sengivukile Momsy.*]
MOMSY: Oh my Lord! [*Hawu Nkosi yami!*]
MAVIYO: Where are we? Where did we sleep? [*Ilapha lana? Silele kuphi?*]
MOMSY: Oh my! Well! It was rough yesterday! Just as I was telling you about your wife Dolly, and that she had taken a lover, you just dropped down – like that! From the pain! I got such a fright I nearly fainted as well ... [*Hhawu! Phela kwenzeke inqaba, lana izolo! Kuthi ngesikhathi ngikuxoxela indaba phela kaDolly, umkakho ukuthi useqomile, wamane waqimuka phansi kabuhlungu kanjani. Lungishaye uvalo ngacishe kwasamina ngaquleka* ...]

Only later in the same episode do listeners meet Maviyo alone, hear his words of pain and regret, and catch him as he weeps and arrives at some degree of self-knowledge. This, perhaps, is the breaking of the husband's body, a physical and symbolic violence, that parallels Dolly's earlier experience. A rag of a man, taken to be a beggar by Momsy and hardly recognisable by his family, his sister-in-law, MaMthembu, almost throws him out of the yard as he approaches his brother's house, before she tackles him about his red eyes:

[8] Table Mountain Messengers' song, 2013 (see Gunner 2014).

MAMTHEMBU: With your eyes as red as that, Junior Father of Mthembeni, well, really, a person could swear you've been crying! [*Indlela abomvu ngayo amehlo akho, Yise Omncane kaMthembeni, hhayi impela, umuntu angafunga ukuthi ubukhala!*]

BHUT´ SIMON: Uh! Hey man! Just please leave, MaMthembu. You can't do anything but mess things up. It's just your way! [*Nxi! Hhayi mani! Awuphume lapha weMaMthembu. Ayikho into eyaziyo wena ngaphandle kokubheda, umthetho wakho!*]

MAMTHEMBU: Oh, I'm so sorry, my dear, I was just throwing it out there. I won't utter another syllable! [*Hawu ngiyaxolisa Baba, sengizowuhlaba inhlali. Ngeke ngisathi vu!*]

The lack of a clear resolution at the end of this powerful 50-episode serial, with its layers of both domestic banter (as above) and serious conflict, is part of the complexity of its message: the drama tells its listeners that it is never easy to resolve the authority and value of custom when it plays out in the wider world. Dolly and Maviyo fail in their 'proper marriage', and, at a number of moments in the episodes, the weight of prejudice against women stands out, and 'respect', *inhlonipho*, in its many shades of language and social practice, is scrutinised. What is the use of ceremony, the weight of the past in the present, if it all fails women? Yet this is not a simple tale. In Episode 33, a changed Dolly, a woman of the world of work, now in a more affluent class, can generously present a 50 rand note to her astonished sister-in-law, MaMthembu, for her to buy food for her own husband, Bhut' Simon, because she has worked too late to cook. Yet Dolly remarks to her lover, Brian, 'They will criticise me, but will they criticise Maviyo?' She uses the analogy of a razor, as opposed to a knife:

DOLLY: You know what? [Criticism] should come in the form of a razor – sharp on two sides, not like a knife, with only one sharp edge. [*Uyaz' ukuthini? Kuyobakuhle uma bezobabukhali 'zindawo zombili njenge 'reyza, bangabibukhali nganxanye njengommese.*]

Within the dense, rich world of the dramas as a body of ongoing narrative assertions, repeats such as *Bakikiza* can be novel dramatic statements, with the power to reshape listener identities as they feed anew into an individual's 'repertoire of listening' (Douglas 1999: 221). Each repeat thus speaks to its new time in a shifting way; the voices within the drama hold a different set of resonances; the temporalities of the drama and the audience are never identical.

Alongside the repeats there are new offerings. Often, these attempt to capture and interpret the topical anxieties of the moment. Still, in some form, the frame of the family remains. One of the two dramas running in November and December 2016, *Umvuzo*, was written by a young writer, Mbali Ngonyana. Her central focus was on the exploitation of young women both by older women, sometimes family members, and by older men. In this drama, the character held up for exposure and ridicule is a headteacher whose predatory advances on a student are carefully dissected for listeners. As a foil, a second young girl is both a school student and an author, determined to write for her peers about the experience of being HIV positive, how it can be avoided, and, if necessary, how it can be lived with. Ngonyana experiments in this second plot strand with father–daughter relationships. She sets up a close and touching relationship between the young girl and her father, who is HIV positive and initially hostile to the idea of his daughter writing about the condition. He feels vulnerable and too much in the public eye, but eventually he realises the value of what she is doing and encourages her.

Umvuzo lacks the playing with language and rich textuality that mark the vocal register of *Bakikiza*, and it features no star performers from an earlier generation of actors. Yet it held its audience[9] and invited engagement with the pressing contemporary questions of women's creativity and agency both within the family and in the arena of health and social issues. The dramas thus continue to create a symbolic space in which to engage with the urgent questions of life and death – both physical and social. The family, seen through the lens of the everyday and as a multivalent symbol, despite being under scrutiny and subject to sharp critique, remains at the core of the dramas' working aesthetic. This, I argue, shows that radio dramas are a genre of the present as well as of the past, that they are engaged in searching for a new centre, a changing imaginary, and are taking their listeners with them in an uncertain and difficult era.

Making listeners work: the *Vuka Mzansi* Breakfast Show[10]

I turn now to ask how a single, key time slot in the life of a station can create a forum, a modern *inkundla* or public meeting place for Ukhozi FM listeners. This is not a new phenomenon: Ukhozi FM's forerunner, Radio Zulu, created a number of forums that replicated earlier or ongoing sites of sociality, and other radio stations had similarly reproduced older sites of African sociality in which specific issues could be aired – sometimes delicate topics touching on sexual matters, or difficult and ignored subjects such as rights and grievances (Mano 2004; Englund 2011). I examine moments of the programme from 2012 to the present that show its anchor presenters in dialogue with listeners, attempting to define and create a contemporary central public space of belonging. It was a space of constant interaction, where a key notion was that belonging was akin to the process of building, and that it took hard work.

[9] Information from Faith Xaba as the drama progressed in November/December 2016.
[10] The section discussing Lucky Sibiya, Mr Magic, appeared first in Gunner (2017).

Although the two anchor presenters on whom I focus – Linda Sibiya, 'Mr Magic', and his successor, Sipho Mbatha, 'Sgqemeza' (Big Head) – were charismatic figures with clearly defined personae, each had an interactive voice in a public space that could be personal at one moment and touch on issues of broad social importance the next. The interactive response of listeners was taken for granted, and at the time of 'Mr Magic' in particular, the give and take between presenter and public was striking. Even the brief but popular slot featuring a motivational talk was far from being a sermon. It was more like a brief tussle between the speaker and the psyches of the countless listeners, focusing at that moment on what needed to be captured and made sense of in their own lives – to help them live.

From 2008 to the end of January 2014, the *Vuka Mzansi* Breakfast Show, which ran from 6.30 a.m. to 9 a.m., was anchored by the charismatic DJ Linda Sibiya, known as Mr Magic to listeners but frequently also addressed by his honorific clan praise name, Sotobe. Like the DJs of American radio in the 1950s, who created their own profiles, personae, vocal styles, and nicknames as they navigated new social and racially diverse communities (Douglas 1999: 230), Mr Magic was a pioneer and a risk-taker. The audience comprised a cross-section of Zulu speakers, some Zimbabwean SiNdebele speakers,[11] and a number of non-first-language speakers, largely from the Gauteng region and the province of KwaZulu-Natal, and they buzzed around the programme like honey bees. If they phoned in, they often spoke as frankly to Sibiya as if he were their brother or close friend, or even a son – and perhaps he was all of these, and much more. Was this the voice of people who had recently become part of the life of the public space? A subject confident of the right to speak in the new era (Kunreuther 2014)?

The *Vuka Mzansi* Breakfast Show drew listeners like a magnet. The programme and its anchor exuded an upbeat optimism, at times raunchy, always edgy, and absolutely in touch with listeners as they called in to urge him on, greet him, or debate with him in a seemingly endless stream of two-way communication.[12] The programme was fast-moving

[11] There are no official figures on the number of Zimbabwean SiNdebele speakers who listen to Ukhozi FM. Random questioning that I have carried out and casual observation in and around Johannesburg suggest that there are many. SiNdebele and isiZulu can be seen as a single language, and are far closer than isiXhosa and isiZulu. This is because of the close historical links between the AmaNdebele of Zimbabwe and AmaZulu of South Africa. Mzilikazi Khumalo, founder of the Matabele nation (*isizwe*) in the early nineteenth century, was a general in King Shaka's army who defected and went north with his band of followers, settling in the vicinity of present-day Bulawayo, in the west of what is now Zimbabwe (see Omer-Cooper 1966; Ranger 2010). There are also long links of migration for work between SiNdebele speakers from the west of Zimbabwe and those from the Johannesburg region, and shared points of culture in music, particularly jazz.

[12] Linda Sibiya, Mr Magic, was dismissed by the SABC on 30 January 2014. There were many rumours surrounding his dismissal, one being that he had been too open in his

and wide-ranging, urgent and inclusive. It was about the quotidian, the grain of people's lives, but it mixed in the extraordinary and the unexpected. Hence, perhaps, Sibiya's nickname, Mr Magic. Sometimes he touched on a topic that lay beneath the layers of what could be shared as public knowledge, because it was too risky or simply too secret. Mr Magic 'had guts', people would say, to even approach such subjects – for instance, the use of secret powers to become wealthy, or the link between snakes and the occult. Listeners dipped in and out of the programme as they travelled in taxis or cars, as they bustled about at home, or as they sat in a bar or a spaza shop, or even on the pavement. Their ears were attuned to the fast-moving show that was held together by the patter and distinctive vocal mannerisms of Linda Sibiya's voice and 'style' (Ferguson 1999), by his larger-than-life radio persona. Thick and fast, the audience interacted with this media personality who was able to make each listener feel more alive, more connected, and able to share moments of their lives with him.

As they did so, the teeming, fast-paced urban and peri-urban morning spaces – particularly in Durban and Johannesburg – became part of the programme. The impression was of listeners on the move, of life charged with the energies of both sorrow and happiness. The scene changed moment by moment as Mr Magic shifted his gaze around his public. Thus, he might directly call on listeners 'at the [taxi] ranks' (*emarankini*) or 'in taxis' (*emateksini*). Sometimes he would bring in the streets and street corners of the town, evoking a scene with a phrase. A particular group could suddenly be pulled in: 'And you men playing *mlaba'laba* [a draughts game played on pavements and street corners], I'm talking to you too.' Many listeners would know that, back in the mid-1990s, Sibiya himself had walked the hard streets of central Johannesburg with no money in his pockets, as a young man, the son of a security guard, hoping to find work at the SABC. Hardship, humble beginnings, ambition, and then success were part of his own narrative (Moncho 2013).[13]

Political questions, particularly those on everyone's lips, featured both obliquely and overtly on *Vuka Mzansi* and were part of Sibiya's drive to reflect the complex feelings of his listening public, which were at a new level of tension because of unrest in the mining industry. Yet he did not neglect other matters of importance to listeners. On 25 June 2012, he tackled the topic of what he termed 'basic education and quality education', a subject of great concern to many listeners, struggling to educate their children in state schools with problems of discipline, drugs, and

support for the governing party, the ANC. However, eventually a charge of fraud was laid against him (see Lekgau 2015).

[13] Mr Magic's life of hardship followed by success was well known to listeners (see Moncho 2013).

poor teaching. He had a lawyer as his invited guest on the programme. The calls came flooding in. As he took them and his guest responded, pulling no punches in her critique of the current inequalities in the South African school system, he remarked, in the form of a riddle or pithy saying: 'The ministers' children train in private schools, but not ours.'

Such a swift, neat creation of an inner public simply through 'not ours' demonstrated Mr Magic's rhetorical skill and provided a brief glimpse of who he saw as his primary audience. At moments such as this, there was a strong sense of a resistant national public outside the circle of elite power. A few days later (on 2 July), the topic of the call-in shifted to the unpredictable and dynamic former ANCYL leader Julius Malema, who was to play a crucial leadership role in the aftermath of the Marikana tragedy the following month. 'Why does Julius Malema get so much airtime? Is he a future ANC leader?' asked Mr Magic.

As the dramatic Lonmin platinum workers' strike held the attention of the country in the months of July and August 2012, in one session Mr Magic turned to 'the topic of freedom and what it means at the present time, almost 20 years since the first democratic elections in 1994'. On Wednesday 16 August, the fateful day when the miners were shot by police at Marikana, leaving 34 miners dead, Sibiya had as his guest on *Vuka Mzansi* the theatre personality, apartheid-era cultural activist, and theatre entrepreneur Mbongeni Ngema. Known to local and world audiences for his struggle musical *Sarafina!*, which toured to much acclaim in Britain and North America in the 1980s, Ngema remains an icon of South African culture, a figure who, through his theatrical work, has helped shape the consciousness of the late apartheid and present era.[14] Mr Magic, with respectful formality, addressed Ngema by his clan praise name, Madlokovu. He then played songs from *Sarafina!*: first, the *Sarafina!* version of the struggle song '*Isizwe 'simnyama*' (The Black Nation), which he followed with 'Freedom Is Coming'. He then asked Mbongeni Ngema, with studied casualness: 'Well, what about "Freedom Is Coming"? What about freedom now?'

Ngema avoided a direct answer, which would lead to a comparison of 'then' and 'now', and the show moved on. Nevertheless, the question struck at the heart of the turmoil and anxiety of the populace as the Marikana events played out on the media and in popular talk. However, not all discussions at the time were on such 'large' topics. The daily events of people's lives and their everyday trials were never far away. An open session on 'problems' had a number of women listeners calling in to

[14] *Sarafina!* played in New York from 1988 to 1989 and in London in 1990. It drew much of its inspiration from redeploying within the musical's narrative the popular and widely circulating struggle songs of the late anti-apartheid era, such as the two recalled here for listeners by Mr Magic (see Gilbert 2007).

talk about the state of taxis; this, again, implicitly points to listeners who were not part of the elite, not car owners, but who made their way across Durban and Johannesburg squeezed into the fast-moving combi-taxis that were an integral part of mobility in the cities and provided a link for many between the city and smaller towns and rural areas all over Africa. '*Amatekisi ayizikorokoro* [The taxis are old and broken down],' said a woman who called in. And no one contradicted her.

Mr Magic's radio persona on *Vuka Mzansi* and on Ukhozi FM came to a sudden and, for many, a heart-wrenching end on 31 January 2013. The SABC dismissed him and he disappeared from the airwaves. But he remained part of people's lives and conversations. There were speculations: *IsoLezwe* (Eye of the Nation), the fast-growing Durban-based Zulu daily, surmised that Mr Magic was seen as being too close to the ruling party, the ANC. It carried a picture of Mr Magic dancing with President Zuma at the latter's birthday party. Other accounts spoke simply of friction with senior SABC management.

Why did people like him so much? As one listener put it almost two years after he had gone, 'Linda Sibiya was a motivator. If you are down he'll pull you up. If you think you're going to die tomorrow, he makes you think "No!"'[15] Sibiya's genius was that, like the fabulous radio men of another uncertain era, the United States in the 1950s on the cusp of the civil rights struggle, he could make risk a virtue (Douglas 1999). Through the *Vuka Mzansi* show in the years from 2008 to early 2014, the condition of living in a state of uncertainty became a manageable and exhilarating mode of being in the world for his radio public, in rural as well as urban areas. Very shortly before he left, he had this to say in an interview:

> Because of my poor background my love for radio grew at a young age because I just wanted a platform to talk to people and tell them they can make it . . . Radio is real and it's not about fame or money. It's about passion on an everyday level. It's about working with time and painting pictures in people's minds.[16]

Linda Sibiya has taken his place in the genealogy of Ukhozi FM's near legendary figures of the past and the present. Although sonically absent from Ukhozi FM, he remains a key figure in its 'translation' from the pre-1990 era into the late post-apartheid, highly mediated time of angst and desire as the station navigates a role for itself and for its publics from many social strata and diverse places.

[15] A man from Soweto – not a first-language speaker of isiZulu – told me this as I interviewed him informally, while he repaired my security system in Melville, Johannesburg, on 2 December 2015. This confirms my hunch that many listeners to Mr Magic in the Gauteng region were multilingual and had isiZulu as one of a bundle of languages they spoke. They tuned in to the *Vuka Mzansi* Breakfast Show out of preference.

[16] See Moncho (2013).

The demise – or at least the disappearance – of Linda Sibiya from *Vuka Mzansi* may have lost the show some listeners, but under its successive anchor, Sipho Mbatha, or 'Sgqemeza', the programme maintained its role as a steady centre from which audiences could try to hold their lives together. The genial successor to the mercurial Linda Sibiya created his own persona and his own public, supported by his co-presenter, Zama Ngcobo. It was during Sgqemeza's time as anchor, on 8 June 2014, that Pastor Zondo made his pronouncement about 'connecting'. In the full text, Pastor Zondo reminded listeners not to forget about 'people left on their own'. Points of contact were important, Zondo urged. 'Dogs can chase a buck together,' he told his listeners. Alone, one could do nothing – and it was implied that, if you leave people to be on their own, something has failed. The ethic of building things together has become one of the hallmarks of Sgqemeza's time at *Vuka Mzansi*. In November/December 2016, he and Zama Ngcobo moved to the provincial northern KwaZulu-Natal town of Ladysmith, to Ezakheni (Build for Yourselves) Plaza, to interact with listeners on the Morning Show, which is now also known as #VMBS. A few days later, Sgqemeza and the #VMBS team were at the provincial capital of Pietermaritzburg. Part of his brief was to round off the station's contribution to the annual national campaign of '16 Days of Activism against Violence to Women and Children', which was due to end on 10 December. Again, there were many voices of people of different ages and from diverse parts of the province speaking on air. The place names flew in: Mpendle, Donnybrook, Camperdown, Mbali, Sobantu, Edendale. The councillor for Greater Edendale came on and said, 'What I like about Ukhozi is that you do not just sit in your offices, you come out and meet us, talk to us.' He talked about the qualities needed in a good councillor, because, as Sgqemeza reminded him, there have been many bad ones. 'You have to be able to listen to people,' the speaker retorted. 'And you have to respect people.' Then Sgqemeza threw in a key question: 'What will you do to try to stop rape from happening?' 'It's we, fathers who have to concentrate on raising our sons properly so that we raise them knowing that they must respect women in the proper way,' the public official replied.

On the same day, the Breakfast Show shifted briefly from Pietermaritzburg, with its struggle to build the present and to overcome its memories of a violent and divided past, to the Moses Mabhida stadium in Durban, where Ukhozi FM and the provincial government were hosting a memorial for the late gospel singer Sifiso Ncwane, who had died suddenly at a young age on 5 December (the very day on which former President Mandela had passed away three years earlier). Ncwane's passing received national coverage and was an example of how popular music, through gifted and loved individuals, can speak to a national public and a national imaginary. The song mentioned most often in tributes to him was *Kulungile Baba* (It Is Good, Father). A caller on the *Vuka Mzansi* show mentioned that he had heard it playing from his neighbour's house as he was about to commit suicide, and it had held him back. Among the dignitaries and luminaries at the memorial service at Moses Mabhida stadium, two familiar names were mentioned: Linda Sibiya, who had a prominent role in the proceedings, and Pastor Zondo, the motivational speaker. We do not hear Linda Sibiya's voice, but we are told that he has asked Pastor Zondo the question, 'Why should a man who loved God go so soon?' Pastor Zondo's reply is simple and profound: 'Every man has his day [to go]; there is no bird which flies but does not come down to earth.'

Pastor Zondo's answer – with its call for acceptance and its reference to the possibilities of 'flight' – can be seen as summing up the ethos of both a programme and a station as they attempt to hold their own with fierce competition from stations such as Metro FM, Radio 702 and others, and to build a site at the centre of people's lives. The present success of the station is not a given, but it absorbs and galvanises its diverse audiences who live as part of the wider national imaginary but draw strength from the energy and resilience provided by Ukhozi FM. A serial radio drama such as *Bakikiza*, which interrogated both modern marriage and 'tradition', and the important *Vuka Mzansi* Breakfast Show, with its probing, play, and inclusivity, give a sense of a radio culture that pushes the boundaries, ready to reinterpret yet remain faithful to the loyalty of its numerous and far-flung listeners who need the station as much as the station needs them. In a way, it is a community of the voice that has maintained its presence through periods of turbulence and suffering. It has carved out an African voice that is both regional and transnational, and, as a station, it has played a key role in defining the never settled black modern.

Conclusion
Dances of power

Isiningi leminyaka Ukhozi senzela okuhle
Ndiza njalo Khozi olumaphikophiko
Uthuthukise umculo wethu
Siyabonga SABC ngokuthuthukisa umculo wethu
Ndiza njalo Khozi olumaphikophiko
Siyabonga niqhubekele phambili.
Ukhozi has done good things for us year after year,
Keep flying Black Eagle with your wide wings
Keep pushing our song forward
Thank you SABC for promoting our song
Thank you, may you go from strength to strength.

General Singers[1]

I have tried in *Radio Soundings: South Africa and the Black Modern* to arrive at a deeper understanding of how radio soundings – and, within that frame, voice itself – work on multiple levels within a station such as Radio Bantu, a tainted space from which Ukhozi FM grew, the largest and arguably the most important radio station in Southern Africa today. I argue that, through the growth of the station and the role of particular broadcasters, Ukhozi FM now has a recognisable political voice. It is not party political; rather, it is a forum, a public space to which politicians often turn in order to be heard. This powerful space within the democratic state, now over 20 years old, without doubt grew out of the early broadcasters' struggle to have a voice and to clear the metaphorical ground for a *black* voice – not at all what the ideologues intended. To a great extent, the success of the station can be said to stem from K. E. Masinga and his early colleague Hubert Sishi. But there was a host of other broadcasters who opened up the path for others. Remarkably, these included a number of women, such as Winnie Mahlangu, Lawrentia Madlala, Lindiwe Ntuli, and Vicky Masuka. As I have pointed out, the often multiple roles broadcasters played gave added weight to their voice,

[1] General Singers, *Cothoza* Group, leader Dumisani Mahlobo, singing in their weekly practice session at George Goch Hostel, 12 August 2016, recorded by Liz Gunner and Dudu Dlamini.

as the layers of association played off each other in the minds of listeners. They were both intimate figures and public personalities, mediated through the sound of the voice. In this way, the alternative, perhaps unnoticed, public sphere for the black mediated voice came into being. As other African-language stations were established, here, too, dominant voices came into play, even though the apartheid regime's gatekeepers were always present, always there to contend with, to duck and deceive through the use of language (Lekgoathi 2009; 2011). The material soundings of broadcast voices are thus complex things, keeping in play these different registers of voice – sometimes dramatist and actor, and sometimes actor and public personality, as was the case with Eric Ngcobo, who played the part of Sigidi in the radio drama *Yiz' Uvalo*, which seemed to mirror the civil war violence in KwaZulu-Natal in the mid-1980s.

The first chapter of this book showed King Edward Masinga setting his own discursive stamp on the radio station to which, it could be argued, he gave birth in 1941 when he strode into the SABC offices in West Street, Durban and asked for – demanded – a slot for Zulu broadcasting. Perhaps he did more than that, and initiated a new 'dance'. K. E. Masinga, too, can be seen as revealing the impact of this convergence of voice. In a sense, he had to fend off versions of a register of voice that the station's gatekeepers wanted to foist upon him, and to make new ones to give voice to wider aspects of black radio that the gatekeepers either were unaware of or wished to obscure. The obituary recording for 'the King', which went out on Ukhozi FM on 21 June 1990, had a certain schizophrenic quality to it. The solemn voices of those who, on some level, were the gatekeepers and represented the old regime largely praised Masinga as an ethnic figure and commented on his involvement in the programme on '*amagugu akwaZulu*' – the treasures of Zululand. Douglas Fuchs, former director general of the Nguni Services, described him as a '[p]roud lover of the Zulu people from which he sprang'. What was stressed by his colleague Hubert Sishi was his involvement in popular culture and his freewheeling song-making, his careless pleasure in strumming his own *maskanda* song on a steel string guitar, a song in which a suspicious husband asks his wife why there is a smell of cigarette smoke in the house. He also remarked on Masinga's love of Frank Sinatra's music and his interest in tap-dancing – something he enthused about on his return from his American tour in 1957. In the middle there was his translation of Shakespeare plays, something that both colleagues and gatekeepers commented on. There was a song for him by the premier *cothoza* group, Ladysmith Black Mambazo. The composer Mzilikazi Khumalo's choral farewell, '*Lala ngokuthula*' (Rest in Peace), was also played. In other words, Masinga's voice, the fulcrum of his complex career and his charisma as originator of a successful Zulu radio station, was an extremely multi-layered one. This, perhaps, sums up the complicated trajectory of the station itself as a radio sounding, through the apartheid decades into the new era. If Masinga loved Sinatra's song 'My Way', the song could also be taken as a marker of the station itself, which forged its own way, bringing together an extraordinary range of styles, memories, and new and old cultural repertoires to create the distinctive voice that the station has come to embody.

In his recent study of Ghana and early radio and film, Peter Bloom (2014) has shown how the BBC's specific broadcasting style in the colonial era represented a voice of authority. Not only was that voice English, but it was also grounded in the accent of RP (received pronunciation) as set down by John Reith, the BBC's first director. Governing at one remove, through an instrument such as radio (or film), makes the voice the medium of power: 'the "voice" of authority is grounded in the ventriloquist's art of governance' (ibid.: 137).[2] Bloom continues to note how 'film and radio embody a tactics of address in the name of government institutions', and how 'radio was understood to be the most pointed instrument of advanced administration by the end of the interwar period' (ibid.: 140–2).

If we consider Masinga and his use of voice, of claiming power through naming, and his emphasis throughout his long career in radio on a distinct kind of performance, we can see an attempt to destabilise the sounds of authority, an act not of mimicry but of substitution. Masinga wished to create a new platform of authority with a language other than English as its centre. Again, Bloom remarks on how a radio report of the 1951 address by Governor Charles Arden-Clarke at the opening of the Gold Coast Legislative Assembly was introduced with 'musical fanfare that might introduce the presence or coming of the King' (2014: 148). This was not, of course, the drum music and language of a Ghanaian figure of authority such as the Asantehene, but that of the British King. So the fanfare served as a sort of signature tune of royalty, a mark of 'state spatiality' (Ferguson and Gupta quoted in Bloom 2014). If we think of sound signifying authority or shifting authority, then we can begin perhaps to see that even the signature tunes of important personages on subaltern radio stations – which is what Radio Bantu and Radio Zulu were – signified alternative and new kinds of authority and presence. Certainly, Masinga, 'the King', had his signature tunes, which Ukhozi FM still plays when it recalls him as one of the station's past heroes.

The mediation of voice and the suggestion of different kinds of embodiment can indeed be seen as discursive markers in the making of alternative narratives of presence – in history and within what was a seemingly unending narrative of colonial presence and continuing

[2] See Brian Larkin's invaluable study *Signal and Noise* (2008).

Figure 9.1 King Edward Masinga, affectionately known as Shobane kaMangethe, the father of Radio Bantu, Radio Zulu, and Ukhozi FM, in full traditional Zulu regalia while being interviewed by WTIC-TV in the US, 1956.
Source: Independent Media.

influence, even after the end of colonialism or imperial rule. So voice in radio can become a means of presenting a different way of seeing and being in the world, and, I suggest, a different, resistant historical narrative – and, in some cases, alternative narratives of the present (Englund 2015b; 2018). It is this idea of voice that has also played out in this book. For instance, the plurality and multiple different voices captured by Lewis Nkosi in his Transcription Centre broadcasts were one of the ways of laying the groundwork for contemplating possible new orders, new regimes of knowledge, and new ways of being in the world at a time of potential shifts of power. Likewise, Bloke Modisane's BBC radio plays were not simply superb plays for Radio 3 or Radio 4 but presented, through voice, something new, where the politics of race and territorial struggles of power and loyalty over three decades played out to home audiences in an increasingly diverse Britain.

Modisane, Nkosi, and Masinga, through their historical interventions, were themselves making the modern; their work as creators, actors, and producers, shaping 'the new', brings to mind Lynn Thomas's observation that 'the modern can be explored "as a discursive tool deployed by historical actors"' (Thomas quoted in Bloom, Miescher, and Manuh 2014: 4). Even the refracted 'Friends, Romans, countrymen' uttered by Caiphas Sedumo, the seemingly impotent intellectual, in the drunken bar in Modisane's first BBC radio drama, *The Quarter Million Boys*, is indeed momentarily the voice of resistance (Chapter 3). And it is this speech from Shakespeare's *Julius Caesar* that is all that remains in the Sound Archives of the SABC of K. E. Masinga's translations of the Bard into his own language of resistance, isiZulu (Chapter 1).

To hone in on the radio drama, then, the presence of multiple uses of voice as the carrier of the endless stream of narratives of the everyday became a way of establishing, through radio, a modern presence that could continue to define both itself and its path. Radio drama marked a huge body of knowledge of the everyday and of the workings of language and possible moralities and histories that would not go away. Instead, the dramas and their listeners and makers became part of a public that may not have been noticed by or have been of interest to the dominant discourse and the visible public sphere, but was nevertheless powerfully present. The medium of radio thus became a resistant medium, and the aesthetic of the dramas was, on some level, itself resistant.

How, then, did this carry over into the new regime? As I discussed in Chapters 7 and 8, the cruelty and erasing violence of the apartheid era were gone. Nevertheless, the power of English as the dominant language of the public sphere – something, perhaps, not entirely foreseen by any of the key actors in the making of the new order – meant that, in some ways, African languages were still seen as being on the outside of the 'present', of the real modern; they were languages for ceremonies of the past and a door into heritage but not public forums for the mix of cultures in the new state. This, we know, was an illusion, but it was a powerful one. The radio stations that broadcast in the nine recognised African languages of

South Africa in the new order became even more important as carriers of the voices of the new publics who worked and had their being and existence, their music and artistic forms – at least some of the time – in isiZulu, SePedi, isiXhosa, and so on. To some extent, a station such as Ukhozi FM had to carry the voices of the important subalterns of the new era, those on the margins of power and wealth, and those not comfortable or 'not entirely at ease' (as Chinua Achebe might say) in English, the dominant language of power, government, and a section of the media. It thus continued as a voice of the present, maintaining its role as coming from a different discursive centre and domesticating a certain multivocal morality of the present (Englund 2018).

What exactly are Ukhozi FM's presence and voice in the second decade of the new millennium? The drama slots are still in place at the same times. I am assured – not by station managers but by people on the ground – that they are still listened to, by those *emalokshini* (in the townships), in small offices, and in workrooms in the cities, and by those in the scattered rural communities of Zulu-speaking areas, which cover KwaZulu-Natal, parts of Mpumalanga, the Eastern Cape, and Ekurhuleni and Gauteng. A recent repeat of an old favourite, the serial drama *Umcebo Owalahlekayo* (The Lost Treasure) – in which K. E. Masinga featured as a father – so agitated listeners that they called in to Lady D's long mid-morning programme to discuss *why* Mandla, the son, seemed to be losing his integrity and turning to crime in his attempt to clear his name.[3] Such levels of radio and listener interaction are not unknown on other radio stations – such as the BBC with its long-running *The Archers* programme – and are also typical of African-language programmes in Malawi and Zimbabwe (Mano 2004; Englund 2011). Here, I point simply to the ongoing potency of the serial dramas alongside listeners' devotion to television soaps and other radio channels. People switch channels and frequencies far more now. The veteran broadcaster Tim Modise remarked at a meeting in Cape Town in 2012 that people are now less loyal to particular stations; just like with football clubs, people no longer have a single loyalty.[4]

A brief look at Ukhozi FM in the present day shows the dynamism of the African-language presence in the country. The station has fast-moving morning primetime shows, such as the *Vuka Mzansi* Breakfast Show, where hosts such Linda Sibiya, 'Mr Magic', or the present anchors, 'Sgqemeza' Mbatha and Zama Ngcobo, attract large audiences and interactive conversations (Chapter 8). So confident are listeners of their place as Ukhozi FM's public that voices from the *skwatta* camps, the informal settlements that cluster densely along the Rand and in many other urban

[3] Personal communication with Faith Xaba, 16 October 2016.
[4] Personal communication with Tim Modise, 26 June 2012.

centres, can phone in confidently to say who they are, where they are from, make their point, and sign off. Music from gospel to rap features. Jazz has its place too, as do the quiet but often probing evening programmes such as *Abasiki bebunda* (Evening Gossip), run by the commentator and anchor Bongani Mavuso. Voice, then, continues to be a major actor in this public-service station of the SABC. Despite complex debacles at the SABC itself, as national political wills impinge on the public broadcaster's code of objectivity, Ukhozi FM goes on.[5] Like other African-language post-apartheid stations, it has built broadcast communities that 'help to create a public language culture and bring into continual contact and creative interaction rural and urban speakers of major regional languages' (Coplan 2011: 148). It mirrors and interacts with a vibrant, multi-layered community, resistant, searching, and engaged. One that is 'irrepressibly self-promoting and multivocal' (ibid.: 148). In all this, I maintain, the 'King', with his dance of power, his voice, and his performativity, played a crucial role. In the decades since the optimistic beginning of the 'new era' (*uhlel' olusha*), Ukhozi FM has reshaped itself in an aggressive, deregulated media marketplace. It has found a way to build on the strong tradition of Zulu-language broadcasting that many producers of culture fashioned over more than 60 years.

How, then, might we compare the figure of the voice and its expressive uses on this station with the radio histories discussed by Fisher in relation to the northern Australian indigenous voice, and with Kunreuther's descriptions of the appearance in the public sphere of the Nepali 'voiced subject' (Kunreuther 2014; Fisher 2016)? The station I have discussed, focusing largely on the important radio genre of *amastori/imidlalo yomoya* or radio drama, has affinities with both but has its own distinctive character and history of voice and soundings. First, like the satellite stations of northern Australia discussed by Fisher, Ukhozi FM – and before it Radio Zulu and Radio Bantu – without doubt became part of a global majority of colour in an important way, a counter-voice to the dominant presence of the old voices of empire and colonialism. What is also relevant to this study about Fisher's work is the notion of healing – 'suturing', as he puts it – which also applies to the history of pain, pleasure, and building (*ukuwakha*) that took place along the continuum of Radio Bantu, Radio Zulu, and Ukhozi FM. As Jacob Dlamini pointed out in his autobiography *Native Nostalgia* (2009), Radio Zulu was part of the pleasure of life in the townships during his youth in the 1980s. It was a support, and a resource. What can we say 'Zulu did to radio?' – a question asked by Achille Mbembe when I presented an early version of

[5] See, for instance, the lead report in *The Star*: 'Faith [Muthambi]: "It wasn't me." Minister [of Communications] stands firm in her denial of wrong doing', *The Star*, 15 December 2016.

Chapter 1 at a seminar at Wiser, Wits University, in 2012? Perhaps most importantly what 'Zulu did to radio' was to open up a space for the black voice, and, in this case, for a specific African language that is widely spoken not only in its home province of KwaZulu-Natal but also on the Witwatersrand and as far as western Zimbabwe. More particularly, it opened up a space for an inclusive presence of a black modernity that defined, painfully at times, what it would take from the past and what it would take with it into the present. As some of the radio dramas show, there was much that could not be jettisoned easily, such as the power of indigenous healers, the dangers of the occult, and the presence of the ancestors in the decisions of the living.

Kunreuther's profound study of the subject moving through the intimacy of autobiography, and particularly love, into the public space, with voice as its central point of analysis, has strong echoes within Radio Zulu and Ukhozi FM. So too, does Harri Englund's new study of voice, creativity and polyphony on a contemporary Zambian radio station (2018). As I set out earlier, I see the intimacies of the dramas as creating a space for the public subject of the post-1990s. Why should love and the family be at the heart of the dramas through the decades? Why should public intimacy create the modern subject in the democratic space? It is the medium of voice, mediatised, that enables this. My study therefore suggests that the speaking voice in the public space has been of huge importance in the very different history of South Africa captured by the *longue durée* and present vigour of the station now called Ukhozi FM.

Bibliography

Abraham, William A. 1962. *The Mind of Africa*. London: Weidenfeld and Nicolson.

Abu-Lughod, Lila. 2000. 'Modern subjects: Egyptian melodrama and post-colonial difference' in T. Mitchell (ed.), *Questions of Modernity*. Minneapolis MN: University of Minnesota Press.

Agamben, Giorgio. 1993. *Homo Sacer, Sovereign Power and Bare Life*. Translated by D. Heller-Roazen. Stanford CA: Stanford University Press.

Aitchison, John. 1989. *The Pietermaritzburg Conflict: experience and analysis*. Pietermaritzburg: Centre for Adult Education, University of Natal.

2010a. 'The Pietermaritzburg conflict: experience and analysis' in J. Aitchison, W. Leeb, and J. Vaughn (eds), *Political Violence in the Natal Midlands: the Unrest Monitoring Project papers 1988–1994*. Pietermaritzburg: University of KwaZulu-Natal.

2010b. 'The civil war in Natal' in J. Aitchison, W. Leeb, and J. Vaughn (eds), *Political Violence in the Natal Midlands: the Unrest Monitoring Project papers 1988–1994*. Pietermaritzburg: University of KwaZulu-Natal.

2010c. 'Mpumalanga: a case study of the Natal violence' in J. Aitchison, W. Leeb, and J. Vaughn (eds), *Political Violence in the Natal Midlands: the Unrest Monitoring Project papers 1988–1994*. Pietermaritzburg: University of KwaZulu-Natal.

2015. *Numbering the Dead: the course and pattern of political violence in the Natal Midlands, 1987–1989*. Pietermaritzburg: Natal Society Foundation Trust.

Aitchison, John, Wendy Leeb, and John Vaughn (eds). 2010. *Political Violence in the Natal Midlands: the Unrest Monitoring Project papers 1988–1994*. Pietermaritzburg: University of KwaZulu-Natal.

Alegi, Peter. 2010a. *Laduma! Soccer, politics and society in South Africa, from its origins to 2010*. Scottsville: University of KwaZulu-Natal Press.

2010b. *African Soccerscapes: how a continent changed the world's game*. Athens OH: Ohio University Press.

Anderson, Benedict. 1991. *Imagined Communities: reflections on the origins and spread of nationalism*. Cambridge: Verso.

Appadurai, Arjun. 1991. 'Global ethnoscapes: notes and queries for a transnational anthropology' in R. Fox (ed.), *Recapturing Anthropology: working in the present*. Santa Fe NM: School of American Research Press.

Apter, Emily. 2005. *The Translation Zone: a new comparative literature*. Princeton NJ: Princeton University Press.

Ash, William. 1985. *The Way to Write Radio Drama*. London: Elm Tree.

Ashforth, Adam. 2005. *Witchcraft, Violence and Democracy in South Africa*. Chicago IL: Chicago University Press.

Attwell, David. 2005. *Rewriting Modernity: studies in black South African literary history*. Athens OH: Ohio University Press.
Bakhtin, Mikhail. 1984. *The Dialogic Imagination: four essays by M. M. Bakhtin*. Edited by Michael Holquist, translated by Caryl Emerson and Michael Holquist. Austin TX: University of Texas Press.
Baldwin, James. 1960. *Nobody Knows My Name: more notes of a native son*. London: Michael Joseph.
1963. *The Fire Next Time*. London: Michael Joseph.
Barber, Karin. 1997. 'Introduction' in K. Barber (ed.), *Readings in African Popular Culture*. London: Routledge.
2000. *The Generation of Plays: Yoruba popular life in theater*. Bloomington IN: Indiana University Press.
2007. *The Anthropology of Texts, Persons and Publics: oral and written cultures in Africa and beyond*. Cambridge: Cambridge University Press.
Bauman, R. and Charles L. Briggs. 2003. *Voices of Modernity: language ideologies and the politics of inequality*. Cambridge: Cambridge University Press.
Beecroft, Alexander. 2015. *An Ecology of World Literature: from antiquity to the present day*. London and New York NY: Verso.
Behrend, Heike. 2002. 'Images of an African modernity: the Likoni Ferry photographers of Mombasa, Kenya' in J.-G. Deutsch, P. Probst, and H. Schmidt (eds), *African Modernities: entangled meanings in current debate*. Oxford: James Currey.
Beinart, William. 2005 [1994]. *Twentieth Century South Africa*. Oxford: Oxford University Press.
Bloom, Peter J. 2014. 'Elocution, Englishness and empire: film and radio in late colonial Ghana' in P. Bloom, S. F. Miescher, and T. Manuh (eds), *Modernization and Spectacle in Africa*. Bloomington IN: Indiana University Press.
2016. 'The language of counterinsurgency in Malaya: dialectical soundscapes of salvage and warfare' in I. Aitken and C. Deprez (eds), *The Colonial Documentary Film in South and South-East Asia*. Edinburgh: Edinburgh University Press.
Bloom, Peter, Stephan F. Miescher, and Takyiwaa Manuh (eds). 2014. *Modernization and Spectacle in Africa*. Bloomington IN: Indiana University Press.
Boehmer, Elleke. 2002. *Empire, the National and the Postcolonial 1890–1920: resistance in interaction*. Oxford: Oxford University Press.
Bonner, Philip and Vusi Ndima. 2008. 'The roots of violence and martial Zuluness on the East Rand' in B. Carton, J. Laband, and J. Sithole (eds), *Zulu Identities: being Zulu past and present*. Pietermaritzburg: University of KwaZulu-Natal Press.
Bonnin, Debby. 2000. 'Claiming spaces, changing places: political violence and women's protests in KwaZulu-Natal', *Journal of Southern African Studies* 26 (2): 301–16.
Borstelmann, Thomas. 2001. *The Cold War and the Color Line: American race relations in the global arena*. Cambridge MA: Harvard University Press.
Brennan, James R. 2010. 'Radio Cairo and the decolonization of East Africa' in C. J. Lee (ed.), *Making a World after Empire: the Bandung Moment and its political afterlives*. Athens OH: University of Ohio Press.
Brooks, Peter. 1976. *The Melodramatic Imagination: Balzac, Henry James, melodrama and the mode of excess*. New Haven CT: Yale University Press.

Brown, Ryan. 2014. *A Native of Nowhere: the life of Nat Nakasa*. Johannesburg: Jacana.

Callaway, Henry. 1868. *Nursery Tales, Traditions and Histories of the Zulus*. London: Trübner and Company.

Campbell, James. 2002. 'The Americanization of South Africa' in A. Offenburger, S. Rosenberg, and C. Saunders (eds), *A South African and American Comparative Reader: the best of Safundi and other selected articles*. New York NY: Safundi.

Canonici, Noverino N. 1988. *Izinganekwane: an anthology of Zulu folktales*. Durban: University of Natal.

Carton, Ben. 1998. '"The new generation . . . jeer at me, saying we are all equal now": impotent African patriarchs, unruly African sons in colonial South Africa' in M. Aguilar (ed.), *Politics of Age and Gerontocracy*. Trenton NJ: Africa World Press.

1999. *Blood from Your Children: colonial origins of African generational conflict in South Africa*. Charlottesville VA: University Press of Virginia.

Carton, Benedict, John Laband, and Jabulani Sithole (eds). 2008. *Zulu Identities: being Zulu, past and present*. New York NY: Columbia University Press.

Carver, Richard. 1995. 'Themes and trends in African broadcasting' in R. Carver and T. Naughton (eds), *Who Rules the Airways? Broadcasting in Africa*. London: Article 19 and Index on Censorship.

Chakrabarty, Dipesh. 2000. *Provincializing Europe: postcolonial thought and historical difference*. Princeton NJ: Princeton University Press.

Chatterjee, Partha. 1993. *The Nation and Its Fragments: colonial and postcolonial histories*. Princeton NJ: Princeton University Press.

Chipkin, Ivor. 2007. *Do South Africans Exist? Nationalism, democracy and the identity of the people*. Pretoria: HSRC Press.

Chiumbu, Sarah and Dina Ligaga. 2013. '"Communities of strangerhoods?": internet, mobile phones and the changing nature of radio cultures in South Africa', *Telematics and Informatics* 30: 242–51.

Cohen-Portheim, Paul. 1935. *The Spirit of London*. London: Batsford.

Coldry, Nick. 2013. *Why Voice Matters*. London: Sage.

Cole, Catherine. 2009. *Performing South Africa's Truth Commission: stages of transition*. Bloomington IN: Indiana University Press.

Cole, Jennifer and Lynn M. Thomas (eds). 2008. *Love in Africa*. Chicago IL: Chicago University Press.

Comaroff, Jean and John L. Comaroff (eds). 1993. *Modernity and Its Malcontents*. Chicago IL: Chicago University Press.

2000. 'Millennial capitalism: first thoughts on a second coming', *Public Culture* 12(2): 291–343.

Coplan, David. 1979. 'The African musician and the development of the Johannesburg entertainment industry, 1900–1960', *Journal of Southern African Studies* 5(2): 135–64.

1985. *In Township Tonight! South Africa's black city music and theatre*. London: Longman.

1995. *In the Time of Cannibals: word music of South Africa's Basotho migrants*. Chicago IL: University of Chicago Press.

2007. *In Township Tonight! Three centuries of South Africa's black music and theatre*. Johannesburg: Jacana.

2011. 'South African radio in a saucepan' in Liz Gunner, Dina Ligaga, and Dumisani Moyo (eds), *Radio in Africa: publics, cultures, communities*. Johannesburg: Wits University Press.

Coupe, Laurence. 1997. *Myth*. London: Routledge.

Couzens, Tim. 1983. 'An introduction to the history of football in South Africa' in B. Bozzolli (ed.), *Town and Countryside in the Transvaal: capitalist penetration and popular response*. Johannesburg: Ravan Press.

1985. *The New African: a study of the life and work of H. I. E. Dhlomo*. Johannesburg: Ravan Press.

Crisell, Andrew. 1986. *Understanding Radio*. London: Methuen.

2004. *Understanding Radio*. Second edition. London: Routledge.

Crystal, David. 1987. *The Cambridge Encyclopaedia of Language*. Cambridge: Cambridge University Press.

Davis, Stephen R. 2011. 'Voices from without: the African National Congress, its radio, its allies and exile, 1960–84' in Liz Gunner, Dina Ligaga, and Dumisani Moyo (eds), *Radio in Africa: publics, cultures, communities*. Johannesburg: Wits University Press.

Dhlomo, Herbert Isaac Ernest. 1985. *Collected Works*. Edited by N. Visser and T. Couzens. Johannesburg: Ravan Press.

Dhlomo, Rolfus Reginald Raymond. 1962 [1939]. *Indlela yaBabi*. Pietermaritzburg: Shuter and Shooter.

Dlamini, Jacob. 2009. *Native Nostalgia*. Johannesburg: Jacana Media.

Doke, C. M. and B. W. Vilakazi. 1972. *Zulu–English Dictionary*. Johannesburg: Witwatersrand University Press.

Douglas, Susan J. 1999. *Listening: radio and the American imagination*. Minneapolis MN: University of Minnesota Press.

Du Bois, William Edward Burghardt. 1996 [1903]. *The Souls of Black of Folk*. New York NY: Dover.

Dubow, Saul. 1989. *Racial Segregation and the Origins of Apartheid in South Africa 1919–1936*. Oxford: Macmillan.

2014. *Apartheid, 1948–1994*. Oxford: Oxford University Press.

Dudziak, Mary L. 2000. *Cold War Civil Rights: race and the image of American democracy*. Princeton NJ: Princeton University Press.

Eade, John. 2000. *Placing London: from imperial capital to global city*. Oxford: Berghahn.

Edwards, Brent Hayes. 2003. *The Practice of Diaspora: literature, translation and the rise of black internationalism*. Cambridge MA: Harvard University Press.

Ellis, Stephen. 1989. 'Tuning in to pavement radio', *African Affairs* 352: 321–30.

Emery, Frank. 1879. *The Red Soldier: letters from the Zulu War*. London: Hodder and Stoughton.

Englund, Harri. 2011. *Human Rights and African Airwaves: mediating equality on the Chichewa radio*. Bloomington IN: Indiana University Press.

2015a. 'Multivocal morality: narrative, sentiment and Zambia's radio grandfathers', *HAU: Journal of Ethnographic Theory* 5(2): 251–73.

2015b. 'Forget the poor: radio kinship and exploited labor in Zambia', *Current Anthropology* 56(11): 137–45.

2018. *Gogo Breeze: Zambia's radio elder and the voices of free speech*. Chicago IL: University of Chicago Press.

Erlmann, Veit. 1991. *African Stars: studies in black South African performance*. Chicago IL: University of Chicago Press.

1996. 'But hope does not kill: black popular music in Durban 1913–1939' in P. Maylem and I. Edwards (eds), *The People's City: African life in twentieth-century Durban*. Pietermaritzburg: University of Natal Press.

Fardon, Richard and Graham Furniss (eds). 2000. *African Broadcast Cultures: radio in transition*. Oxford: James Currey.

Farred, Grant. 1996. 'You can go home again. You just can't stay: Stuart Hall and the Caribbean diaspora', *Research in African Literatures* 27(4): 28–48.

(ed.). 1997. *Rethinking C. L. R. James*. Oxford: Blackwell.

Ferguson, James. 1999. *Expectations of Modernity: myths and meanings of urban life on the Zambian Copperbelt*. California CA: University of California Press.

2006. *Global Shadows: Africa in the neoliberal world order*. Durham NC: Duke University Press.

Ferguson, James and Akhil Gupta. 2002. 'Spatializing states: toward an ethnography of neoliberal governmentality', *American Ethnologist* 28(4): 981–1002.

Fisher, Daniel. 2016. *The Voice and Its Doubles: media and music in northern Australia*. Durham NC: Duke University Press.

Foucault, Michel. 1978. *The History of Sexuality*. Translated by R. Hurley. New York NY: Pantheon.

2008. *The Birth of Biopolitics: lectures at the Collège de France, 1978–1979*. Basingstoke: Palgrave Macmillan.

Fryer, Peter. 1984. *Staying Power: the history of black people in Britain*. London: Pluto.

Gautier, Ana María Ochoa. 2014. *Aurality: listening and knowledge in nineteenth-century Colombia*. Durham NC: Duke University Press.

Gellner, Ernest. 1983. *Nations and Nationalism*. New York NY: Cornell University Press.

Gerhart, Gail M. 1978. *Black Power in South Africa: the evolution of an ideology*. Vol. 19. Berkeley CA: University of California Press.

Geschiere, Peter. 2000 [1997]. *The Modernity of Witchcraft: politics and the occult in postcolonial Africa*. Charlottesville VA: Virginia University Press.

Giddy, Patrick and Markus Detterbeck. 2005. 'Questions regarding tradition and modernity in contemporary choral practice (Makwaya)', *Transformation* 59: 26–44.

Gilbert, Shirli. 2007. 'Singing against apartheid: ANC cultural groups and the international anti-apartheid struggle', *Journal of Southern African Studies* 33 (2): 420–41.

Gilroy, Paul. 1993. *The Black Atlantic: modernity and double consciousness*. London: Verso.

Glissant, Eduoard. 1985. *The Ripening*. London: Heinemann.

Gqibitole, Khaya Michael. 2002. 'Contestations of tradition in Xhosa radio drama under apartheid', *English Studies in Africa* 45(2): 33–46.

2012. 'Creativity or control? A study of selected Xhosa radio plays in the apartheid years'. Doctoral thesis, University of KwaZulu Natal.

Grant, E. W. 1929. 'The *izibongo* of the Zulu chiefs', *Bantu Studies* 3(3): 203–44.

Gunner, Liz. 1979. 'Songs of innocence and experience: Zulu women as composers and performers of *izibongo* Zulu praise poetry', *Research in African Literatures* 10(2): 239–67.

1994. *Politics and Performance: theatre, poetry and song in Southern Africa.* Johannesburg: Witwatersrand University Press.

2000a. 'Wrestling with the present, beckoning to the past: contemporary Zulu radio drama', *Journal of Southern African Studies* 26(2): 223–37.

2000b. 'Zulu radio drama' in S. Nuttall and C.-A. Michael (eds), *Senses of Culture: South African culture studies.* Cape Town: Oxford University Press.

2002. 'Resistant medium: the voices of Zulu radio drama in the 1970s', *Theatre Research International* 27(3): 259–74.

2004. 'Africa and orality' in A. Irele and S. Gikandi (eds), *The Cambridge History of African and Caribbean Literature.* Vol. 1. Cambridge: Cambridge University Press.

2005. 'Contaminations: BBC radio and the black artist – Lewis Nkosi's "The Trial" and "We Can't All Be Martin Luther King"' in Lindy Stiebel and Liz Gunner (eds), *Still Beating the Drum.* Amsterdam: Rodopi.

2011. 'IsiZulu radio drama and the modern subject: restless identities in South Africa in the 1970s' in Liz Gunner, Dina Ligaga, and Dumisani Moyo (eds), *Radio in Africa: publics, cultures, communities.* Johannesburg: Wits University Press.

2014. 'Soft masculinities, *isicathamiya* and radio', *Journal of Southern African Studies* 40(2): 343–60.

2017. 'The man from where? Ukhozi FM and new identities in radio in South Africa', *Journal of African Media Studies* 9(1): 61–75.

Gunner, L. and I. Gunner. 2008. '"Where's it gone freedom?" Composing *isicathamiya* in post-apartheid South Africa in the age of 9/11' in B. Carton, J. Laband, and J. Sithole (eds), *Zulu Identities: being Zulu past and present.* Pietermaritzburg: University of KwaZulu-Natal Press.

Gunner, Liz and Mafika Gwala (edited and translated). 1991. *Musho! Zulu popular praises.* East Lansing MI: Michigan State University Press.

Gunner, Liz, Dina Ligaga, and Dumisani Moyo (eds). 2011. *Radio in Africa: publics, cultures, communities.* Johannesburg: Wits University Press.

Guthrie, Malcolm. 1948. *The Classification of the Bantu Languages.* London: Oxford University Press.

Hall, Stuart. 1996. 'The question of cultural identity' in S. Hall, D. Held, D. Hubert and K. Thompson (eds), *Modernity: An Introduction to Modern Societies*, Cambridge: Polity Press.

Hamm, Charles, 1991. '"The constant companion of man": Separate development, Radio Bantu and music', *Popular Music* 10(2): 147–73.

1995. *Putting Popular Music in Its Place.* Cambridge: Cambridge University Press.

Hannerz, Ulf. 1997. 'Sophiatown: the view from afar' in K. Barber (ed.), *Readings in African Popular Culture.* Oxford: James Currey.

Hašek, Jaroslav. 2005 [1930]. *The Good Soldier Schweik.* London: Penguin.

Hemingway, Ernest. 1932. *Death in the Afternoon.* New York NY: Scribner's.

Hendy, David. 2000. *Radio in the Global Age.* Cambridge: Polity Press.

Hirschkind, Charles. 2006. *The Ethical Soundscape: cassette sermons and Islamic counterpublics.* New York NY: Columbia University Press.

Hofmeyr, Isabel. 1987. 'Building a nation with words: Afrikaans language, literature and ethnic identity, 1902–1924' in S. Marks and S. Trapido (eds), *The Politics of Race, Class and Nationalism in Twentieth-Century South Africa.* Harlow: Longman.

1993. *'We Spend Our Years as a Tale That Is Told': oral historical narrative in a South African chiefdom*. London: James Currey.

2003. *The Portable Bunyan: a transnational history of* The Pilgrim's Progress. Princeton NJ: Princeton University Press.

Horrell, Muriel. 1958. 'Race classification in South Africa: its effects on human beings'. Fact paper 2. Johannesburg: South African Institute of Race Relations.

Hughes, Heather. 2011. *First President: a life of John L. Dube, founding president of the ANC*. Johannesburg: Jacana.

Hughes, Langston (ed.). 1961. *An African Treasury: selected by Langston Hughes*. New York NY: Pyramid.

(ed.). 1963. *Poems from Black Africa*. Bloomington IN: Indiana University Press.

2002a [1961]. 'Black Nativity: a gospel song and play for a variable cast' in Christopher C. De Santis (ed.), *Gospel Plays, Operas and Late Dramatic Works*. Vol. 6. Columbia MO: University of Missouri Press.

2002b. *The Collected Works of Langston Hughes: essays on art, race, politics and world affairs*. Vol. 9. Edited by Christopher C. De Santis. Columbia MO: University of Missouri Press.

Hull, Elizabeth and Deborah James. 2012. 'Introduction: popular economies in South Africa', *Africa* 82(1): 1–19.

Hunter, Mark. 2010. *Love in the Time of Aids: inequality, gender and rights in South Africa*. Bloomington IN: Indiana University Press.

Hutcheon, Linda and Mario J. Valdes (eds). 2002. *Rethinking Literary History: a dialogue on theory*. Oxford: Oxford University Press.

Hyslop, Jonathan. 1993. 'A destruction coming in: Bantu Education as response to social crisis' in P. Bonner, P. Delius, and D. Posel (eds), *Apartheid's Genesis 1935–1962*. Johannesburg: Ravan Press.

IBA.1995. *Report on the Protection and Viability of Public Broadcasting Services; Cross Media Control of Broadcasting Services; Local Television Content and South African Music*. Johannesburg: Independent Broadcasting Authority (IBA).

James, Cyril Lionel Robert. 2003 [1932]. *Letters from London*. Oxford: Signal Books.

James, Deborah. 2012. 'Money-go-round: personal encounters of wealth, aspiration and indebtedness', *Africa* 82(1): 20–40.

Jameson, Fredric. 1986. 'Third World literature in the era of multinational capitalism', *Social Text* 15: 65–88.

Joseph, Rosemary. 1983. 'Zulu women's music', *African Music* 6(3): 53–89.

1987. 'Zulu women's bow songs: ruminations on love', *Bulletin of the School of Oriental and African Studies* 50(1): 90–119.

Karis, Thomas and Gail Gerhart. 1997. *From Protest to Challenge: a documentary history of African politics in South Africa, 1882–1990. Vol. 5: Nadir and resurgence, 1964–1979*. Bloomington IN: Indiana University Press.

Kavanagh, Robert. 1981. *South African People's Plays*. London: Heinemann.

1986. *Theatre and Cultural Struggle in South Africa*. London: Zed Books.

Knipe, Michael. 1998. 'Obituary: from tradesman's daughter to stage queen', *Mail & Guardian*, 21–27 August.

Kofsky, Frank. 2001 [1970]. *John Coltrane and the Jazz Revolution of the 1960s*. Expanded and revised second edition of *Black Nationalism and the Revolution in Music*. New York NY: Pathfinder.

Krog, Antje. 1998. *Country of My Skull: guilt, sorrow and the limits of forgiveness in the new South Africa*. New York NY: Random House.

Kruger, Loren. 1999. *The Drama of South Africa: plays, pageants and publics since 1910*. London: Routledge.

2013. *Imagining the Edgy City: writing, performing and building Johannesburg*. Cape Town: Oxford University Press.

Kuan-Hsing, Chen and David Morley (eds). 1996. *Stuart Hall: critical dialogues in cultural studies*. London: Routledge.

Kunreuther, Laura. 2014. *Voicing Subjects: public intimacy and mediation in Kathmandu*. Berkeley CA: University of California Press.

La Hausse de Lalouvière, Paul. 1996. 'The struggle for the city: alcohol, the *ematsheni* and popular culture in Durban, 1902–1936' in P. Maylem and I. Edwards (eds), *The People's City: African life in twentieth-century Durban*. Pietermaritzburg: University of Natal Press.

1989. 'The message of the warriors: the ICU, the labouring poor and the making of a popular political culture in Durban, 1925–1930' in Philip Bonner, Isabel Hofmeyr, Deborah James, and Tom Lodge (eds), *Holding Their Ground: class, locality and culture in 19th and 20th century South Africa*. Johannesburg: Raven Press.

2000. *Restless Identities: signatures of nationalism, Zulu ethnicity and history in the lives of Petros Lamula (c.1881–1948) and Lymon Maling (1889–c.1936)*. Pietermaritzburg: University of Natal Press.

Lamming, George. 1960. *The Pleasures of Exile*. London: Longman.

Lamula, Petros. 1924. *UZulukaMalandela: a most practical and concise compendium of African history combined with genealogy, chronology, geography and biography*. Durban: Star Printing Works.

Larkin, Brian. 2008. *Signal and Noise: media, infrastructure, and urban culture in Nigeria*. Durham NC and London: Duke University Press.

Lekgau, Mashoto. 2015. 'Linda Sibiya on bail for fraud', *Sowetan Live*, 28 February, www.sowetanlive.co.za/entertainment/2015/02/28/linda-sibiya-on-bail-for-fraud (accessed 15 November 2016).

Lekgoathi, Sekibakiba Peter. 2009. '"You are listening to Radio Lebowa of the South African Broadcasting Corporation": vernacular radio, Bantustan identity, and listenership, 1960–1994', *Journal of Southern African Studies* 35 (3): 575–94.

2010. 'The African National Congress's Radio Freedom and its audiences in apartheid South Africa 1963-1991', *Journal of African Media Studies* 2(2): 139–53.

2011. 'Bantustan identity, censorship and subversion on Northern Sotho Radio under apartheid, 1960s to 1980s' in Liz Gunner, Dina Ligaga, and Dumisani Moyo (eds), *Radio in Africa: publics, cultures, communities*. Johannesburg: Wits University Press.

Ligaga, Dina. 2011. 'Radio theatre: the moral play in the historical context of state control and censorship of broadcasting in Kenya' in Liz Gunner, Dina Ligaga, and Dumisani Moyo (eds), *radio in Africa: publics, cultures, communities*. Johannesburg: Wits University Press.

Lutz, Catherine and Lila Abu-Lughod. 1990. 'Introduction: emotion, discourse and the politics of everyday life' in C. Lutz and L. Abu-Lughod (eds), *Language and the Politics of Emotion*. Cambridge: Cambridge University Press.

Makeba, Miriam and James Hall. 1988. *Makeba: my story*. London: Bloomsbury.

Mano, Winston. 2004. 'Renegotiating tradition on Radio Zimbabwe', *Media, Culture and Society* 26(3): 315–36.

McDonald, Peter D. 2009. *The Literature Police: apartheid censorship and its cultural consequences*. Oxford: Oxford University Press.

McLeod, John. 2004. *Postcolonial London: rewriting the metropolis*. London: Routledge.

Meintjes, Louise. 2003. *Sound of Africa! Making music Zulu in a South African studio*. Durham NC: Duke University Press.

Meyer, Birgit. 1999. 'Popular Ghanaian cinema and the "African heritage"'. Working paper 7. The Hague: WOTRO Project, Globalization and the Construction of Communal Identities.

Meyer, Birgit and Annelies Moors. 2006. *Religion, Media and the Public Sphere*. Bloomington IN: Indiana University Press.

Meyer, Birgit and Peter Pels (eds). 2003. *Magic and Modernity: interfaces of revelation and concealment*. Stanford CA: Stanford University Press.

Mhlambi, Innocentia J. 2012. *Popular Culture: IsiZulu fiction and black television dramas*. Johannesburg: Wits University Press.

Mhlambi, Thokozani Ndumiso. 2008. 'The early years of black radio broadcasting in south Africa: a critical reflection on the making of Ukhozi FM'. MA thesis, University of Cape Town.

2015. 'Early radio broadcasting in South Africa: culture, modernity and technology'. PhD thesis, University of Cape Town.

Mitchell, Timothy. 2000. 'Introduction' in T. Mitchell (ed.), *Questions of Modernity*. Minneapolis MN: University of Minnesota Press.

Modisane, Bloke. 1962. 'The rage that racism bred: short story writing in black South Africa', *Negro Digest* 11(12): 61–71.

1986 [1963]. *Blame Me on History*. London: Thames and Hudson.

Modisane, Litheko. 2013. *South Africa's Renegade Reels: the making and public lives of black-centred films*. New York NY: Palgrave Macmillan.

Moncho, Kgomotso. 2013. 'Up close and personal with Ukhozi FM's Mr Magic', RadioBiz, 22 January, http://radiobiz.co.za/2013/01/22/up-close-personal-with-ukhozi-fms-mr-magic/ (accessed 24 November 2015).

Moodie, Dunbar and Vivienne Ndatshe. 1994. *Going for Gold: men, mines, and migration*. Berkeley CA: University of California Press.

Moorman, Marissa. 2008. *Intonations: a social history of music and nation, Luanda, Angola, 1945 to recent times*. Athens OH: Ohio University Press.

2011. 'Airing the politics of nation: radio in Angola, past and present' in Liz Gunner, Dina Ligaga, and Dumisani Moyo (eds), *Radio in Africa: publics, cultures, communities*. Johannesburg: Wits University Press.

Msomi, Welcome. 1996 [1970]. *uMabatha*. Pretoria: Via Afrika/Skotaville.

Mtshali, Oswald. 1975. 'Amagoduka' in Wole Soyinka (ed.), *Poems of Black Africa*. London: Heinemann African Writers Series.

NAC. 1941. *Report of the Native Affairs Commission 1939–1940*. Vol. 99, Part V. Pretoria: Government Printers.

Naipaul, Vidiadhar Surajprasad. 1988 [1963]. *Mr Stone and the Knights Companion*. London: Penguin.

Naude, N. J. 1970. 'Broadcasting' in D. S. Potgieter (ed.), *Standard Encyclopaedia of Southern Africa*. Vol. 2. Cape Town: Nasou.

Ndebele, Njabulo. 1994. *South African Literature and Culture: the rediscovery of the ordinary*. Manchester: Manchester University Press.

Ndlovu, Musawenkosi. 2011. 'The meaning of post-apartheid Zulu media', *South African Journal for Communication Theory and Research* 37(2): 268–90.

Ndlovu, Thabisani. 2013. 'Fixing families through television?', *Cultural Studies* 27(3): 379–403.

Ngubane, Harriet. 1986. 'The predicament of the sinister healer: some observations on "ritual murder" and the professional role of the *inyanga*' in M. Last and G. Cahvanduka (eds), *The Professionalisation of African Medicine*. Manchester: Manchester University Press and International African Institute.

Nienaber, Petrus Johannes. 1966. *Beeld van'n Digter: N. P. van Wyk Louw*. Cape Town: Nasionale Boekhandel.

Nixon, Rob. 1994. *Homelands, Harlem and Hollywood: South African culture and the world beyond*. New York NY: Routledge.

Nkosi, Lethiwe. 2014. 'Black women in post-apartheid South Africa, nation-building and radio: the case of Ukhozi FM'. MPhil thesis, University of Cape Town.

Nkosi, Lewis. 1965. *Home and Exile*. London: Longman.

Nuttall, Sarah. 2009. *Entanglement: literary and cultural reflections on post-apartheid*. Johannesburg: Wits University Press.

Nuttall, Sarah and Cheryl-Ann Michael. 2000. 'Autobiographical acts' in S. Nuttall and C. Michael (eds), *Senses of Culture*. Cape Town: Oxford University Press.

Nyamnjoh, Francis B. 2005. *Africa's Media: democracy and the politics of belonging*. Pretoria: UNISA Press.

Nyembezi, Cyril Lincoln Sibusiso. 1961. *Inkinsela yaseMgungundlovu*. Pietermaritzburg: Shuter and Shooter.

Nyembezi, Cyril Lincoln Sibusiso and Otty E. H. Nxumalo. 1966. *Inqolobane Yesizwe*. Pietermaritzburg: Shuter and Shooter.

Omer-Cooper, John D. 1966. *The Zulu Aftermath: a nineteenth-century revolution in Bantu Africa*. London: Longman.

Patel, Essop (ed.). 1985. *The World of Can Themba*. Johannesburg: Ravan Press.

Paton, Alan. 1948. *Cry the Beloved Country*. London: Scribner.

Peterson, Bhekizizwe. 1994. 'Apartheid and the political imagination in black South African theatre' in L. Gunner (ed.), *Politics and Performance: theatre, poetry and song in Southern Africa*. Johannesburg: Wits University Press.

2000. *Monarchs, Missionaries and African Intellectuals: African theatre and the unmaking of the colonial mind*. Oxford: James Currey.

Pheiffer, P. H. 1982. *Inleiding: N. P. van Wyk Louw. Drie Hoorspele*. Johannesburg: Perskor.

Plaatje, Sol. 1916. *Native Life in South Africa*. London: P. S. King.

Posel, Deborah. 1991. *The Making of Apartheid 1948–1961*. Oxford: Clarendon.

2010 'Races to consume: revisiting South Africa's history of race, consumption and the struggle for freedom', *Ethnic and Racial Studies* 33(2): 157–75.

Ranger, Terence. 2010. *Bulawayo Burning: the social history of a Southern African city 1893–1960*. Woodbridge: Boydell and Brewer.

Reyher, Rebecca Hourwich. 1999. *Zulu Woman: the life story of Christina Sibiya*. Introduction by Christina Sibiya and afterword by Elizabeth Gunner. New York NY: Feminist Press at CUNY.

Rosenthal, Eric. 1974. *You Have Been Listening . . .: a history of the early days of radio transmissions in South Africa*. Cape Town: Purnell.
Rycroft, David. 1975. 'The Zulu ballad of Nomagundwane', *African Language Studies* XVI: 61–92.
1975–6. 'The Zulu bow songs of Princess Magogo', *African Music* 5(4): 41–97.
SABC. n.d. *The Voice – The Vision: the sixty-year history of the SABC*. Johannesburg: Auckland Graphics.
Said, Edward W. 1993. *Culture and Imperialism*. New York NY: Vintage.
Salkey, Andrew (ed.). 1960. *West Indian Stories*. London: Faber and Faber.
Samuel, Raphael (ed.). 1981. *People's History and Socialist Theory*. London: Routledge and Kegan Paul.
Sanders, Mark. 2002. *Complicities: the intellectual and apartheid*. Durham NC: Duke University Press.
Scheub, Harold. 1975. *The Xhosa 'ntsomi'*. Oxford: Oxford University Press and Clarendon Press.
Scott, Sally. 1999. 'A way with words', *Daily News*, 23 August.
Selvon, Sam. 1956. *The Lonely Londoners*. London: Longman.
Shear, Keith. 2012. 'Tested loyalties: police and politics in South Africa, 1939–63', *Journal of African History* 53(2): 173–93.
Sibiya, Dumisani. 2001a. 'Performance to print and print to performance: plays, adaptations and book readings on South African radio', *Journal of African Cultural Studies* 14(2): 137–52.
2001b. 'Media and the development and promotion of African-language literature: Ukhozi FM and Ilanga'. Master's thesis, University of the Witwatersrand.
Sishi, Hubert. 2000. *Imilando yakwaZul: imidlalo yomsakazo* [Histories of Zululand: radio plays]. Edited by D. B. Z. Ntuli. Pretoria: Unisa Press.
Sithole, Jabulani. 2008. 'Changing meanings of the Battle of Ncome and images of King Dingane in twentieth-century South Africa' in B. Carton, J. Laband, and J. Sithole (eds), *Zulu Identities: being Zulu past and present*. Pietermaritzburg: University of KwaZulu-Natal Press.
Soyinka, Wole. 1962. *The Lion and the Jewel*. Oxford: Oxford University Press.
Spitulnik, Debra. 1998. 'Mediated modernities: encounters with the electronic in Zambia', *Visual Anthropology Review* 14(2): 63–84.
2000. 'Documenting radio culture as lived experience: reception studies and the mobile machine in Zambia' in R. Fardon and G. Furniss (eds), *African Broadcast Cultures: radio in transition*. Oxford: James Currey.
2002. 'Mobile machines and fluid audiences: rethinking reception through Zambian radio culture' in F. Ginsberg, L. Abu-Lughod, and B. Larkin (eds), *Media Worlds: anthropology on new terrain*. Berkeley CA: University of California Press.
Stein, Pippa and Ruth Jacobson (eds). 1986. *Sophiatown Speaks*. Johannesburg: Junction Ave Press.
Stein, Sylvester. 1999. *Who Killed Mr 'Drum'?* Cape Town: Mayibuye.
Sterne, Jonathan. 2003. *The Audible Past: cultural origins of sound reproduction*. Durham NC: Duke University Press.
2011. 'The theology of sound: a critique of orality', *Canadian Journal of Communication* 36(2): 207–25.
2012. *The Sound Studies Reader*. London: Routledge.

Steyn, Jacob Cornelius. 1998. Van Wyk Louw,'n Lewensverhaal. *Vol. 2*. Cape Town: Tafelberg.

Stoller, Paul. 1984. 'Sound in Songhay cultural experience', *American Ethnologist* 11(3): 559–70.

Suso, Bamba and Banna Kanute. 1999. *Sunjata*. Translated and annotated by Gordon Innes and Bakari Sidibe in 1974. Edited with new introduction by Graham Furniss and Lucy Duran. London: Penguin Classics.

Switzer, Les. 1985. *Media and Dependency: a case study of the press and the Ciskei homeland*. Athens OH: University of Ohio Press.

Taylor, Charles. 1989. *Sources of the Self: the making of modern identity*. Cambridge MA: Harvard University Press.

Tettey, Wisdom J. 2011. 'Talk radio and politics in Ghana: exploring civic and (un)civil discourse in the public sphere' in Liz Gunner, Dina Ligaga, and Dumisani Moyo (eds), *Radio in Africa: publics, cultures, communities*. Johannesburg: Wits University Press.

Thomas, Lynn. 2011. 'Modernity's failings, political claims, and intermediate concepts', *American Historical Review* 116(3): 727–40.

2008. 'Love, sex and the modern girl in 1930s Southern Africa' in Jennifer Cole and Lynn M. Thomas (eds), *Love in Africa*. Chicago IL: University of Chicago Press.

Titlestad, Michael. 2004. *Making the Changes: jazz in South African literature and reportage*. Pretoria: Unisa Press.

Tomaselli, Ruth, Keyan Tomaselli, and Johan Muller (eds). 1989. *Currents of Power: state broadcasting in South Africa*. Bellville: Anthropos.

Tracey, Hugh and King Edward Masinga. 1945. *Chief Above and Chief Below*. Pietermaritzburg: Shuter and Shooter.

Turrell, Rob. 2001. '*Muti* ritual murder in Natal: from chiefs to commoners (1900–1930)', *South African Historical Journal* 44: 21–39.

Viljoen, Shaun. 2007. 'Proclamations and silences: "race", self-fashioning and sexuality in the trans-atlantic correspondence between Langston Hughes and Richard Rive', *Social Dynamics: A Journal of African Studies* 33(2): 105–22.

Vinson, Robert. 2012. *The Americans Are Coming! Dreams of African-American liberation in segregationist South Africa*. Athens OH: Ohio University Press.

von Eschen, Penny. 1997. *Race against Empire: black Americans and anticolonialism*. New York NY: Cornell University Press.

Warner, Marina. 1994. *Six Myths of Our Time: little angels, little monsters, beautiful beasts and more*. New York NY: Vintage.

Warner, Michael. 2002. 'Publics and counter-publics', *Public Culture* 14(1): 49–90.

White, Hylton. 2012. 'A post-Fordist ethnicity: insecurity, authority, and identity in South Africa', *Anthropological Quarterly* 85(2): 397–428.

Willems, Wendy. 2013. 'Participation – in what? radio, convergence and the corporate logic of audience input through new media in Zambia', *Telematics and Informatics* 30: 223–31.

Willems, Wendy and Winston Mano (eds). 2016. *Everyday Media Culture in Africa: audiences and users*. London: Routledge.

Williams, Raymond. 1989. *The Politics of Modernism: against the new conformists*. London: Verso.

Wood, Felicity. 2003. 'Snakes, spells, Cadillacs and Kruger millions: the weaving of story in oral accounts of the extraordinary life of Khotso Sethuntsa'. Working paper 54. East London: Fort Hare Institute of Social and Economic Research.
Wood, Felicity and Michael Lewis. 2007. *The Extraordinary Khotso: millionaire medicine man from Lusikisiki*. Johannesburg: Jacana Media.

Archives

American archives

Harry Ransom Humanities Research Center, Transcription Centre (HRHRC TC) Archive, University of Texas, Austin.

British archives

African Writers' Club tapes, Sound Archives, British Library, London.
BBC Written Archives, Caversham, Reading.
National Sound Archives, British Library, London.

South African archives

Centre for African Literary Studies (CALS), University of KwaZulu-Natal, Pietermaritzburg.
National English Literary Museum (NELM), Rhodes University, Grahamstown.
SABC Sound Archives.
State Archives.

Archival references

BBC Written Archives. Bloke Modisane files, Caversham, Reading, UK.
1969. *The Quarter Million Boys* (script) by Bloke Modisane, ref. DM795H. Caversham.
SABC Sound Archives. 1942. 'First broadcast to natives', 1/63 (42), 29 September. Auckland Park, Johannesburg.
1955a. '*Zashonaphi Izihlabani*' [Where have all the brave ones gone?], Cat. T (Z/94) 23. K. E. Masinga Road, Durban.
1955b. 'Julius Caesar', E86/86. Auckland Park, Johannesburg.
1958a. 'Malcolm Fleming and K. E. M. Masinga', Acc. 201420, E86/97, 24 January. Auckland Park, Johannesburg.
1958b. 'Malcolm Fleming and K. E. M. Masinga', Acc. 57113, Cat. T (Z/94) 23, 21 January. K. E. Masinga Road, Durban.
1986. 'Izihlabani Zomsakazo' uMnu K. E. Masinga', Acc. 65982, Cat. T (Z/94) 161. K. E. Masinga Road, Durban.

Interviews

Patrick Masonto Buthelezi, with Liz Gunner and Wiseman Masango, Durban SABC, 25 June 1999.
Thami Buthelezi, with Liz Gunner, Pietermaritzburg, 26 June 2012.
Koos Hadebe, with Liz Gunner, Johannesburg, 4 June 1999 and 18 November 2000.
Mandla Langa, with Liz Gunner, Pietermaritzburg, 5 September 2001.
Winnie Mahlangu, phone interview with Liz Gunner, Johannesburg–Durban, 11 August 2006.
Tinyiko Maluleke, with Liz Gunner, Scottsville, Pietermaritzburg, 20 October, 2000.
Khaba Mkhize, with Liz Gunner, SABC Durban, 4 June 1999.
V. V. O. Mkhize, with Liz Gunner, Johannesburg, 21 August 2009.
Bheki Msane and Dumisani Nkosi, with Liz Gunner, SABC Durban, 9 September 1998.
Eric Ngcobo, with Liz Gunner and Wiseman Masango, Durban, 24 May 2000.
with Liz Gunner, Durban, 3 June 2011.
Thando Ngubane, with Liz Gunner and Wiseman Masango, Mkhambathini, 17 November 2002.
Thulas Nkabinde, with Liz Gunner, SABC Auckland Park, 8 March 2007.
Lewis Nkosi, with Liz Gunner, Melville, Johannesburg, 20 October 2007.
Nakanjani Sibiya, with Liz Gunner, Durban, 3 June 2011.
Frikkie Wallis and Schalk van Wyk, with Liz Gunner, SABC Auckland Park, 21 July 2006.

SABC Sound Archives radio dramas

Abangane Ababi [Bad Friends]. 1979. Written by Abigail Zondi, Cat. TS (Z90) 522–57.
Bakikiza Buthule [Wedding Dances in the Silence]. 2000–1. Written by Morris V. Bhengu.
Buzani kuBaba [Ask Father]. 1965. Translated from the isiXhosa radio drama, *Buzani kuBawo* by Winnie Mahlangu, Acc. 24615.
Ebuhleni Nasebubini [For Better and for Worse]. 1999. Written by Dumisani Nkosi, Cat. TS (Z/93) 3100–33.
Ibuzwa Kwabaphambili [The Elders Know Best]. 1965. Written by Alexius Buthelezi (on the life of Johann Strauss), Acc. 71398.
Icala Lombango [The Succession Dispute]. 1967. Written by Muntu Xulu and produced by Bhekinkosi Ntuli (D. B. Z. Ntuli), Cat. TS (Z/67) 89–103.
Impic' Abadala [A Tricky Situation]. 1999. Written by Cecilia Mkhwanazi.
Inkinsela YaseMgungundlovu [The Flashy Man from Pietermaritzburg]. 1963. Written by Sibusiso Nyembezi, Cat. TS (Z/64) 101–37.
Isiboshwa [The Prisoner]. 2000. Written by Mbuso Langa, Cat. TS (Z00) 211–37.
Isiduphunga [The Fool]. 1964. Written by Hubert Sishi, Acc. 24672, Cat. TS (64) 1–41.
Isikhumba Sebhubesi [The Lion's Skin]. 1968. Written by Hubert Sishi, Acc. 24675. Cat. TS (Z/68) 1–80.

Khumbula uDeliwe [Remember Deliwe]. 1964. Written by Mandla Sibiya, Acc. 24623.

Kungenxa Yakho [It's Your Fault]. 1999–2000. Written by Lovemore Wiseman Ndlovu, Cat. TS (Z/00) 211–37.

KwaNhliziyo Ngiyise [Where My Heart Takes Me]. Text recorded by Liz Gunner and transcribed by Wiseman Masango, 1999, but no archival record available in the SABC Sound Archives in Durban.

Ngakubona Ngakubeletha [Suffering! I Know You Well!]. 1999. Written by Innocent Masinga, Cat. TS (Z/99) 176–204.

Raka. 1963. Written by N. P. van Wyk Louw, translated into isiZulu by unknown, musical producer Alexius Buthelezi, Cat. TM 8065(93).

uBhekifa [Look After Your Inheritance]. 1970. Adapted by P. V. Dlamini from the novel by Muntu Xulu, produced by Alexius Buthelezi, Cat. TS (Z/90) 822–61.

Ubongilinda Mzikayifani [You Must Wait for Me, Mzikayifani]. 1974. Produced and written by Bhekisisa Kunene, Cat. (A/90) 183–202.

uChakijane neSalukazi [Chakijane and the Old Woman]. 1975. Written by Alexius Buthelezi, Cat. TD (Z01) 11.

uDeliwe [Contented One]. 1964. Written by Mandla Sibiya, Acc. 24618.

uFakazi [The Witness]. 1999. Written by M. C. Banda, Cat. TS (Z/99) 227–320.

uKubuya KukaNsakansaka eDayimane [Trouble Returns to Diamond Town]. 1996. Written and produced by Alexius Buthelezi, based on Muntu Xulu's novel, starring Winnie Mahlangu and Mary Nontolwane, Acc. 24624.

Umakhandamahlanu [Five Heads]. 1976. Written by Alexius Buthelezi, Acc. 141081, Cat. TD (Z/00) 6.

Umbango Wamawele [The Battle of the Twins]. 1968. Written by Tracey Mdlala and produced by Guybon Mpanza, Cat. TS (Z/70) 1–35.

Umcebo Owalahlekayo [The Lost Treasure]. 1968. Written by D. B. Z. Ntuli, featuring K. E. Masinga, Acc. 24648, Cat. TS (Z/68) 81–122.

Umvuzo [The Reward]. 2016. Written by Mbali Ngonyana.

uNobathakathi. 1963. Written by Alexius Buthelezi, Acc. 40749/43645, Cat. TD(93) 12 and TM8068 (93).

uNokhwezi [Morning Star]. 1964–5. Written by Alexius Buthelezi, Acc. 43640, Cat. CDR (2010) 103 and TM (93) 8066–7.

Yiz' Uvalo and *Umanqob' Isibindi* [In Spite of Fear and The Victor Is Courage]. 1986–7. Written by Morris V. Bhengu, *Yiz' Uvalo*: Cat. TS (Z/93) 1938–77.

BBC radio dramas

Leave Well Alone. 1976. Written by Bloke Modisane. Broadcast on Radio 4 as the 'Saturday Night Theatre', 9 March.

Many Brave Hearts. 1973. Written by Donald McNutt Douglas and adapted for radio by Bloke Modisane. Broadcast on Radio 4 as the 'Saturday Night Theatre', February.

The Quarter Million Boys. 26 September 1969. Written by Bloke Modisane and produced by Charles Lefeaux. First aired on the 'Third Programme' (later Radio 3).

Sitting Duck. 11 August 1980. Written by Bloke Modisane and produced by Brian Wright. Broadcast on Radio 4 as the 'Monday Play'.

The Somebody. 13 July 1987. Written by Bloke Modisane. Broadcast on Radio 3 as the 'Monday Play'.

The Trial. 1969. Written by Lewis Nkosi.

We Can't All Be Martin Luther King. 1971. Written by Lewis Nkosi.

Index

Titles in the series

60. MARK HUNTER *Race for Education: Gender, White Tone, and Schooling in South Africa*
59. LIZ GUNNER *Radio Soundings: South Africa and the Black Modern*
58. JESSICA JOHNSON *In Search of Gender Justice: Rights and Relationships in Matrilineal Malawi*
57. JASON SUMICH *The Middle Class in Mozambique: The State and the Politics of Transformation in Southern Africa*
56. JOSÉ-MARÍA MUÑOZ *Doing Business in Cameroon: An Anatomy of Economic Governance*
55. JENNIFER DIGGINS *Coastal Sierra Leone: Materiality and the Unseen in Maritime West Africa*
54. HANNAH HOECHNER *Quranic Schools in Northern Nigeria: Everyday Experiences of Youth, Faith, and Poverty*
53. HOLLY PORTER *After Rape: Violence, Justice, and Social Harmony in Uganda*
52. ALEXANDER THURSTON *Salafism in Nigeria: Islam, Preaching, and Politics*
51. ANDREW BANK *Pioneers of the Field: South Africa's Women Anthropologists*
50. MAXIM BOLT *Zimbabwe's Migrants and South Africa's Border Farms: The Roots of Impermanence*
49. MEERA VENKATACHALAM *Slavery, Memory and Religion in Southeastern Ghana, c.1850–Present*
48. DEREK PETERSON, KODZO GAVUA, and CIRAJ RASSOOL (eds) *The Politics of Heritage in Africa: Economies, Histories, and Infrastructures*
47. ILANA VAN WYK *The Universal Church of the Kingdom of God in South Africa: A Church of Strangers*
46. JOEL CABRITA *Text and Authority in the South African Nazaretha Church*
45. MARLOES JANSON *Islam, Youth, and Modernity in the Gambia: The Tablighi Jama'at*
44. ANDREW BANK and LESLIE J. BANK (eds) *Inside African Anthropology: Monica Wilson and her Interpreters*
43. ISAK NIEHAUS *Witchcraft and a Life in the New South Africa*
42. FRASER G. MCNEILL *AIDS, Politics, and Music in South Africa*
41. KRIJN PETERS *War and the Crisis of Youth in Sierra Leone*
40. INSA NOLTE *Obafemi Awolowo and the Making of Remo: The Local Politics of a Nigerian Nationalist*
39. BEN JONES *Beyond the State in Rural Uganda*
38. RAMON SARRÓ *The Politics of Religious Change on the Upper Guinea Coast: Iconoclasm Done and Undone*
37. CHARLES GORE *Art, Performance and Ritual in Benin City*
36. FERDINAND DE JONG *Masquerades of Modernity: Power and Secrecy in Casamance, Senegal*
35. KAI KRESSE *Philosophising in Mombasa: Knowledge, Islam and Intellectual Practice on the Swahili Coast*
34. DAVID PRATTEN *The Man-Leopard Murders: History and Society in Colonial Nigeria*
33. CAROLA LENTZ *Ethnicity and the Making of History in Northern Ghana*

32. BENJAMIN F. SOARES *Islam and the Prayer Economy: History and Authority in a Malian Town*
31. COLIN MURRAY and PETER SANDERS *Medicine Murder in Colonial Lesotho: The Anatomy of a Moral Crisis*
30. R. M. DILLEY *Islamic and Caste Knowledge Practices among Haalpulaar'en in Senegal: Between Mosque and Termite Mound*
29. BELINDA BOZZOLI *Theatres of Struggle and the End of Apartheid*
28. ELISHA RENNE *Population and Progress in a Yoruba Town*
27. ANTHONY SIMPSON *'Half-London' in Zambia: Contested Identities in a Catholic Mission School*
26. HARRI ENGLUND *From War to Peace on the Mozambique–Malawi Borderland*
25. T. C. MCCASKIE *Asante Identities: History and Modernity in an African Village 1850–1950*
24. JANET BUJRA *Serving Class: Masculinity and the Feminisation of Domestic Service in Tanzania*
23. CHRISTOPHER O. DAVIS *Death in Abeyance: Illness and Therapy among the Tabwa of Central Africa*
22. DEBORAH JAMES *Songs of the Women Migrants: Performance and Identity in South Africa*
21. BIRGIT MEYER *Translating the Devil: Religion and Modernity among the Ewe in Ghana*
20. DAVID MAXWELL *Christians and Chiefs in Zimbabwe: A Social History of the Hwesa People c.1870s–1990s*
19. FIONA D. MACKENZIE *Land, Ecology and Resistance in Kenya, 1880–1952*
18. JANE I. GUYER *An African Niche Economy: Farming to Feed Ibadan, 1968–88*
17. PHILIP BURNHAM *The Politics of Cultural Difference in Northern Cameroon*
16. GRAHAM FURNISS *Poetry, Prose and Popular Culture in Hausa*
15. C. BAWA YAMBA *Permanent Pilgrims: The Role of Pilgrimage in the Lives of West African Muslims in Sudan*
14. TOM FORREST *The Advance of African Capital: The Growth of Nigerian Private Enterprise*
13. MELISSA LEACH *Rainforest Relations: Gender and Resource Use among the Mende of Gola, Sierra Leone*
12. ISAAC NCUBE MAZONDE *Ranching and Enterprise in Eastern Botswana: A Case Study of Black and White Farmers*
11. G. S. EADES *Strangers and Traders: Yoruba Migrants, Markets and the State in Northern Ghana*
10. COLIN MURRAY *Black Mountain: Land, Class and Power in the Eastern Orange Free State, 1880s to 1980s*
9. RICHARD WERBNER *Tears of the Dead: The Social Biography of an African Family*
8. RICHARD FARDON *Between God, the Dead and the Wild: Chamba Interpretations of Religion and Ritual*
7. KARIN BARBER *I Could Speak Until Tomorrow:* Oriki, *Women and the Past in a Yoruba Town*
6. SUZETTE HEALD *Controlling Anger: The Sociology of Gisu Violence*

5. GUNTHER SCHLEE *Identities on the Move: Clanship and Pastoralism in Northern Kenya*
4. JOHAN POTTIER *Migrants No More: Settlement and Survival in Mambwe Villages, Zambia*
3. PAUL SPENCER *The Maasai of Matapato: A Study of Rituals of Rebellion*
2. JANE I. GUYER (ed.) *Feeding African Cities: Essays in Social History*
1. SANDRA T. BARNES *Patrons and Power: Creating a Political Community in Metropolitan Lagos*